Thorium-Powered Abundance

Fuel for Bold Space Exploration, Sustainable Energy, Plentiful Freshwater, Thriving Agriculture

Michael Lee Anderson

Published by INOV8R Press LLC
Derry, New Hampshire, USA
https://inov8r.com

ISBN (Paperback): 979-8-9992688-1-5
ISBN (eBook): 979-8-9992688-0-8
ISBN (Hardcover): 979-8-9992688-2-2
ISBN (Audiobook): 979-8-9992688-3-9
Cover design and graphic figures by Rav Astra
Printed in the United States of America

First Edition
10 9 8 7 6 5 4 3 2 1

Contents

Help With the Acronyms   7
Introduction: Why Water? Why Thorium?   11

Part One
Part I: Parched Earth, Powered Future

1. THE PROMISE OF THORIUM   15
    A Different Nuclear Path   15
    From Water Crisis to Desalination Revolution   17
    Not a Silver Bullet, but a Strong Contender   18
    Kirk Sorensen's Vision: TED Talk Insights   20
    Conclusion   24

    Endnotes — Chapter 1   25

2. WATER SCARCITY AND THORIUM-POWERED
    DESALINATION   27
    The Global Strain on Water Systems   28
    Reverse Osmosis: Water Independence at
    Guantanamo Bay   31
    Water Desalination: How it Works   33
    Brine Management: Why Thorium is a Breakthrough
    for Desalination   36
    Implementing Nuclear-powered Desalination: Case
    Examples   39
    Tech in Context: Where Thorium Fits in the
    Nuclear Innovation Landscape   43
    Summary: A Future of Clean Water, Clean Energy   46

    Endnotes — Chapter 2   49

Part Two
Part II: Innovation Ignited

3. INNOVATION AND THE THORIUM
   BREAKTHROUGH   53
   What Makes the Thorium Cycle Different   54
   Energy Efficiency and Safety   56
   Why Thorium Was Overlooked   58
   The Thorium Cycle: Smaller, Safer, More Efficient
   Reactors   59
   Summary: What Still Needs to Happen - From
   Challenge to Momentum   60

   Endnotes — Chapter 3   63

4. THE REACTOR BLUEPRINT: COMPARING
   TECHNOLOGIES FOR A CLEAN ENERGY ERA   65
   Reactor Fuels: What We Burn for Power   66
   Coolants: Keeping the Core Stable   67
   Moderators: Slowing Neutrons for a Chain Reaction   70
   Reactor Types: How the Pieces Fit Together   72
   SMRs: Small Reactors, Big Promise, With Lessons
   Still Unfolding   79

   Endnotes — Chapter 4   83

5. POWERFUL POSSIBILITIES: A CALL FOR
   INNOVATORS AND INVESTORS ACROSS
   INDUSTRY NEEDS   85
   The Oxygen of Novelty: Emerging Technologies   87
   The Heat of the Market Pull: Where's the Demand   91
   How to Fuel it with Funding   101
   The Catalysts: Who are the Innovators?   115

   Endnotes — Chapter 5   125

Part Three
Part III: Fixing the Grid, Feeding the World

6. POWER WHERE IT'S NEEDED: WATER AND FUEL   133

The Water-Energy Nexus   133

Responsive Power: Thorium for a Dynamic Grid   136

Turning Seawater into Synthetic Fuel: Seafineries
and the Hydrogen-Carbon Transition   137

Endnotes — Chapter 6   143

7. PEACE THROUGH POWER   144

Reimagining Global Energy Security   144

Strength of Local Power: Thorium's Global Edge   145

Cooling Conflict: Reducing Tensions Through
Energy Abundance   148

Case Studies in Clean Power: Regional Leaders and
First Movers   148

Peace Through Power: A New Vision for
International Collaboration   150

Empowering Remote Communities with Microgrids   154

Smart Policy for Smarter Power   155

How You Can Help: Grassroots to Global Action   158

Turning Policy into Power: US Reforms Are on
the Way   160

US Nuclear Power Insurance and the Regulatory
Straitjacket   162

US Executive Orders and the Thorium Opportunity   164

The Way Forward: A New Architecture for Peace   167

Endnotes — Chapter 7   171

8. THRIVING AGRICULTURE   174

A Showcase for Thorium's Environmental Promise   174

Farming the Future: Thorium and Clean Agriculture   175

Thorium is Rewriting the Story of Nuclear Waste
and Nuclear Safety   182

Building Trust in Clean Power   184

Endnotes — Chapter 8   187

Part Four
Part IV: The Long Horizon

9. THORIUM SYNERGY IN THE FUTURE OF FUSION   191
    Fusion's Promise...and Its Obstacles   191
    Thorium MSRs: The Practical Path Forward   192
    Fusion and Fission: A Complement...Or Crossroads?   193
    A Shared Technology Base   194
    A Multi-Solution Reactor Grid   195
    Toward a Hybrid Energy Future   196

    Endnotes — Chapter 9   201

10. POWERING THE FINAL FRONTIER: THORIUM MSRS AND THE ARCHITECTURE OF SPACE CIVILIZATION   202
    The Challenge   202
    Step 1: Development: Proving Power in the Void   203
    Step 2: The Moon Colony: Establishing a Foothold   206
    Step 3: The Journey to Mars: Propulsion for the Next Frontier   209
    Step 4: Mars: Energy for a Second Home   211
    Step 5: Deep Space Probes: Power Beyond Sunlight   213
    Step 6: Mining and Refueling: Infrastructure Among the Asteroids   214
    Step 7: Autonomy: Systems That Think, Repair, and Adapt   217
    Step 8: Peace Through Power: A Shared Infrastructure for a Shared Future   218
    Summary: Powering the Final Frontier   222

    Conclusion: A Future Worth Building   227
    Additional Reading – and Watching   235
    Bibliography   241
    Glossary of Acronyms and Abbreviations   251
    Appendix: A Technology Companion for Chapter 10: Powering the Final Frontier: Thorium MSRs and the Architecture of Space Civilization   257
    Acknowledgments   305

Help With the Acronyms

Here's a quick guide to key terms:

Reactors

MSR - Molten Salt Reactor: Uses liquid fluoride salts for cooling and fuel, safer and more efficient than traditional reactors.

LFTR - Liquid Fluoride Thorium Reactor: A type of MSR that uses thorium, offering abundant energy and less waste.

SMR - Small Modular Reactor: Compact reactors that can be factory-built and transported easily, ideal for remote locations.

Desalination:

RO - Reverse Osmosis: A common method that removes salt from seawater using a semi-permeable membrane.

MED - Multi-Effect Distillation: Uses multiple evaporation and condensation stages.

MSF - Multi-Stage Flash: Employs flash evaporation chambers for desalination.

Nuclear fuel cycles:

U-235 - Uranium-235: A natural isotope used in reactors, known for high energy but produces long-lived waste.

Pu-239 - Plutonium-239: A man-made isotope used in breeder reactors that can create more fuel than it consumes.

Th-232 - Thorium-232: A naturally abundant isotope used in thorium reactors, transforming into U-233 (Uranium-233) for energy.

Organizations:

IAEA - International Atomic Energy Agency: Promotes safe nuclear energy use worldwide.

TEA - Thorium Energy Alliance: Advocates for thorium nuclear technology.

IDA - International Desalination Association: Promotes desalination innovation and water reuse.

NIA - Nuclear Innovation Alliance: Focuses on advancing nuclear technology development.

A full glossary of acronyms is available at the end of the book and online at https://inov8r.com.

"We are now examining in the United States today the mixed economic-technical question of whether very large scale nuclear reactors can produce unexpected savings in the simultaneous desalination of water and the generation of electricity.
We will have before this decade is out or sooner a tremendous nuclear reactor which makes electricity and at the same time gets fresh water from salt water at a very competitive price.
What a difference this can make to the United States. And indeed, not only the US but all around the globe where there are so many deserts on the ocean's edge."
- President John F. Kennedy, June 21, 1961
Speech delivered for the inauguration of the Freeport Texas desalination plant. Kennedy, J. F. (1961, June 21). White House Press Office.

Introduction: Why Water? Why Thorium?

Author's remarks:

When I was eighteen, I served on a US Navy submarine. It was a self-contained world beneath the waves powered by a compact nuclear reactor. My job was auxiliary systems: the oxygen we breathed, the water we drank, and all the habitation systems for survival. We were isolated deep below the surface, yet we never worried about running out of air or water. That self-sufficiency stayed with me. If we could achieve this undersea, what could we do to solve resource problems ashore?

Years later, while deployed on a Navy Spruance-class destroyer, I experienced a critical water shortage. Our obsolete flash distillation system was supposed to turn seawater into drinkable water but failed us. Its temperamental salinity sensors dumped perfectly good water back into the ocean, mistaking it for contamination. Our ship went without a reliable source of fresh water for two long weeks. Rationing became our reality. Showers were banned, laundry was postponed, and our morale tanked.

Relief came when a newer destroyer pulled alongside and replenished our fresh water through a hose connection. As the other ship's crew ribbed us, splashing in a kiddie pool on their flight deck and staging mock showers with soap bars, our Executive Officer lashed out at us for not being able to produce the volumes of water that they could. What, then, was their water distillation advantage? Reverse osmosis (RO). Their system forced seawater through membranes to filter out the salt, delivering pure water nearly as fast as they could pump it.

That day, I saw firsthand the life-saving potential of water tech. If reverse osmosis could save a ship, what other benefits could it have for water shortages worldwide? However, I also understood the limit; RO systems need a lot of energy. And that energy usually comes from fossil fuels. Today's systems burn carbon to get water, trading one crisis for another. If large-scale ocean desalination plants are figuratively converting diesel to water, that is not affordable or sustainable in the long run. Renewables like solar and wind help but are intermittent and often insufficient for large-scale desalination.

So, what if we paired the efficiency of RO with a clean, abundant energy source? What if we could unlock the oceans for coastal cities, drought-stricken farms, and growing populations without accelerating climate change?

That's when I found thorium.

Part One

Part I: Parched Earth, Powered Future

Chapter 1

The Promise of Thorium

A Different Nuclear Path

T horium isn't new. Scientists have known about it for over a century, and it's been studied seriously since the Cold War.[1] But it never stepped into the spotlight. Uranium-powered reactors took over, largely because they supported both electricity and weapons development.[2] Thorium, on the other hand, offered something quieter: safer, cleaner energy with little potential for weaponization.[3]

In particular, thorium works well when used in molten salt reactors (MSRs).[4] Instead of solid fuel rods, these reactors dissolve thorium in a liquid fluoride salt that circulates through the system with the consistency of a hot salty liquid.

This design has several major advantages:

- **No high pressure**: Unlike water-cooled reactors, MSRs don't operate under extreme pressure, reducing risks and plant size by up to 80%.

In MSRs, thorium fuel is dissolved in a stable fluoride salt that circulates at atmospheric pressure, eliminating the thick-walled vessels and massive containments that dominate light-water plants. [5]

- **Passive safety**: If temperatures rise beyond design limits, a frozen "salt plug" in a drain line at the base of the reactor melts, the liquid fuel falls into a shielded tank, and the chain reaction stop without pumps or operator action.[6]
- **Waste reduction**: Thorium produces far less long-lived radioactive waste compared to uranium.[7]
- **Proliferation resistance**: Thorium's byproducts cannot be readily weaponized.[8] That is one reason why it has been rejected by nations looking for both simultaneous energy and weapons capabilities because, unlike uranium and plutonium, thorium lacks a direct path for weapons use.

Liquid Fluoride Molten Salt Thorium Reactor

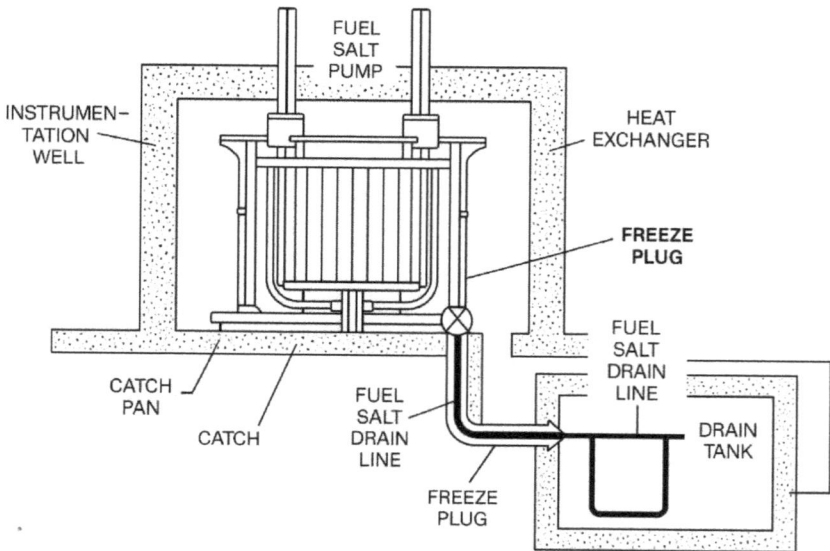

Abundance adds further advantages. Thorium is estimated to be three to four times more common than uranium in the Earth's crust, and it is already co-mined with rare-earth elements in India, Australia, and the American West.[9] Studies from Oak Ridge and the IAEA suggest that the world's electricity needs for a single year could be supplied by the thorium recovered from an area the size of a football field.[10] A lifetime supply could fit in the palm of your hand.[11]

Yet despite all this, thorium development stalled in the 1970s. The reasons were political, not technical.[12] Uranium infrastructure was already in place. Cold War strategy focused on weapons-grade fuel. Thorium simply didn't fit the plan.

But the world has changed.

Water access, grid resilience, and climate goals are priorities today.[13] The timing is right to revisit what thorium can do, especially when paired with modern challenges like desalination and agriculture.

From Water Crisis to Desalination Revolution

More than two billion people already live in water-stressed regions.[14] By 2050, that number may exceed five billion.[15] As climate change worsens droughts and floods, water security becomes central to national security, food production, and economic stability.

Global Freshwater Stress Indicators by Region

Region	Water Stress Level	Primary Stress Factors	Population Affected	Energy-Related Implications
Middle East & North Africa	Extremely High	Scarcity, aquifer depletion	400+ million	Heavy desalination use, grid pressure
Sub-Saharan Africa	High	Infrastructure gaps, climate variability	300+ million	Diesel dependence, weak grids
South Asia	High	Monsoon variability, groundwater depletion	1.6 billion	Irrigation energy demand
Western United States	Moderate to High	Drought, aquifer depletion	100+ million	Competing urban/ag demand, grid flexibility
Northern China	High	Industrial overuse, seasonal river variability	600+ million	Reliance on coal power, water-for-energy conflict
Southeast Asia	Moderate	Monsoon variability, limited storage	300+ million	Hydropower risk, urban expansion stress
Northern Europe	Low	Stable climate, strong infrastructure	200+ million	Low stress, hydropower-dominant
Australia	High	Aridity, salinity, infrastructure strain	25 million	Heavy use of desalination, distributed microgrids

Desalination can help, but only if we reduce how much energy it uses. Most desalination plants today run on fossil fuels.[16] That creates a sustainability paradox: we burn fuel to make water, which accelerates the warming that causes drought in the first place.

Pairing RO with thorium offers a practical solution.[17] A molten salt reactor can deliver continuous, carbon-free energy and waste heat, making it suitable for high-throughput desalination and improving seawater preheating efficiency. One thorium-powered plant could meet a city's water needs while also supplying electricity to the grid.

Unlike solar or wind, which fluctuate, thorium offers a steady, reliable flow of energy.[18] It's also more compact and easier to scale than traditional uranium plants, especially when deployed as small modular reactors (SMRs).[19] These could be built in factories, shipped to the site, and activated quickly, which would be a game-changer for coastal areas and island nations facing water shortages.

For example, Australia has considered thorium for both energy and desalination, given its vast thorium reserves and arid interior.[20] Saudi Arabia and the UAE, which rely on desalination for survival, have launched research into thorium-powered reactors as part of their long-term sustainability plans.[21] And in the US, discussions are growing around integrating advanced nuclear into climate-resilient infrastructure.[22]

Not a Silver Bullet, but a Strong Contender

Let's be clear: thorium isn't a cure-all. It doesn't replace solar or wind. It complements them.

Renewables are excellent when the sun shines and the wind blows. But desalination systems, cities, and data centers need baseline power 24/7. Grid batteries are improving, but they're material-intensive and presently so costly that getting a short-term payback or even a longer-term return on investment is challenging.[23] Conventional uranium

SMRs are ahead in licensing, but they still struggle with some of the traditional nuclear challenges, specifically large footprint and waste handling.

In contrast, Thorium MSRs offer a compelling balance of highly efficient, safe, reliable energy, along with a small land footprint and low environmental impact.

Of course, it won't be easy. Regulatory inertia, supply chain development, and public misunderstanding about nuclear safety remain real hurdles.[24] But the tide is turning. And with population growth and resource depletion, the stakes for water, food, and energy have never been higher.

The purpose of this book is to invite innovators, investors, policy leaders, and the public to consider thorium with fresh eyes. This book is filled with insightful ideas about what can be achieved, along with real-life examples of innovators who are leading the way.

The technology isn't decades away. Prototypes exist. Licensing frameworks are forming. Momentum is building.

Timeline of Thorium Advocacy: From Weinberg to Sorensen

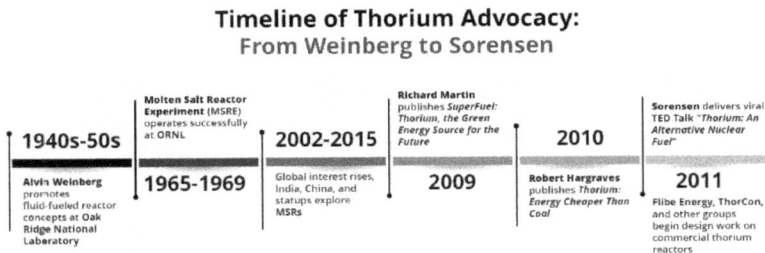

Timeline of Thorium Advocacy:
From Weinberg to Sorensen

1940s-50s	Molten Salt Reactor Experiment (MSRE) operates successfully at ORNL	2002-2015	Richard Martin publishes *SuperFuel: Thorium, the Green Energy Source for the Future*	2010	Sorensen delivers viral TED Talk "*Thorium: An Alternative Nuclear Fuel*"
Alvin Weinberg promotes fluid-fueled reactor concepts at Oak Ridge National Laboratory	1965-1969	Global interest rises, India, China, and startups explore MSRs	2009	Robert Hargraves publishes *Thorium: Energy Cheaper Than Coal*	2011
					Flibe Energy, ThorCon, and other groups begin design work on commercial thorium reactors

Kirk Sorensen's Vision: TED Talk Insights

Author's remarks:

My interest in thorium began with a TED Talk by Kirk Sorensen, a name that has since become synonymous with the push for liquid fluoride thorium reactors.[25] Watching him speak, I was captivated by the idea that a small sphere of thorium, costing less than $150, could supply an individual's lifetime energy needs. That moment was a revelation. Sorensen's talk illuminated the vast potential of thorium, an element more abundant than uranium, found deep in the Earth's mantle and prevalent across the Rocky Mountain states. His words resonated with me, sparking a curiosity that would later guide my professional endeavors.

The practicality of Sorensen's vision became clear when I was working as a systems engineer at an aerospace company, especially as I considered the challenges faced by military operations in conflict zones like Afghanistan. In those environments, logistical convoys, often targeted by insurgents, became the lifeline for forward operating bases, delivering crucial supplies such as fuel and water. Tragically, more than a third of US troop fatalities during these deployments were linked to these vulnerable supply lines.[26] This stark reality compelled me to explore how technology could reduce such risks. I led a study on the potential of small modular thorium reactors to power forward bases, thereby reducing dependency on hazardous convoys. The idea was to leverage thorium's potential to provide a safe, stable, and independent energy source, ultimately helping to save lives.

Meeting Kirk Sorensen was a pivotal moment. I asked to meet him for lunch in Huntsville, Alabama. I found myself face-to-face with the man who had inspired my newfound interest in thorium. I asked him about the possibility of fielding small modular mobile reactors at forward bases. Sorensen spoke candidly about the promise and potential of thorium technology and didn't shy away from acknowledging the challenges. He explained that the technology was still in development, slowed by public policy hurdles and the need for substantial investment.

Despite these obstacles, Sorensen's enthusiasm was contagious, reaffirming my commitment to pursue this path. I recognized that solutions were years away, which was accurate since this meeting occurred in 2013. I followed up with the research arm of a defense agency, and they asked about safety, which is a natural concern for thorium molten salt reactors. Indeed, Sorensen's advocacy for thorium is rooted in its inherent safety and efficiency.[27]

In his TED Talk, Kirk Sorensen highlighted thorium reactors as a cleaner, more efficient form of nuclear technology. He focused on molten salt reactors, which use liquid fluoride salts as both fuel and coolant, improving thermal efficiency and safety. Sorensen's talk sparked considerable interest among the public and scientific communities, leading to extensive discussions about the feasibility of thorium as a long-term, mainstream energy source.[28]

This ongoing dialogue challenges traditional views on nuclear energy and reevaluates thorium's potential role in our energy future. By integrating thorium with renewable energy sources, countries can reduce their carbon emissions and reliance on fossil fuels. This shift also offers geopolitical advantages, allowing nations to secure their energy needs independently from volatile global markets.[29]

Sorensen's TED Talk presents a strong case for thorium, but some nuclear advocates still question its short-term feasibility for a few key reasons. First, the existing nuclear industry has heavily invested in uranium-based fuel cycles, creating resistance to change. Second, thorium reactors, particularly liquid-fueled molten salt designs, face significant engineering challenges that need new regulations, materials testing, and safety validations, which all require time and funding. Lastly, there is little commercial experience with thorium, leading governments and utilities to be hesitant to adopt a technology that lacks large-scale operational proof, despite its potential benefits. These factors hinder its adoption, even as interest in thorium grows in academic and policy circles.

Through his involvement with Flibe Energy, a leader in thorium reactor development, Sorensen is committed to advancing this technology. Flibe Energy focuses on creating liquid-fueled thorium reactors that operate without uranium enrichment, aiming for zero carbon emissions and a sustainable energy future.[30] I, too, can envision a world powered by clean, abundant thorium energy and advocate for its adoption. Meeting with Sorensen reinforced my belief in thorium's potential as a game-changer.

1.5 Flibe Energy and Molten Salt Reactors

Flibe Energy is an inspiring example of innovation in the ever-evolving field of nuclear energy.[31] Their mission is the development of thorium-based molten salt reactors. This technology promises to redefine how we produce and consume energy. Flibe Energy is not just about creating reactors; it's about creating a sustainable future where energy is abundant, clean, and safe. Their focus is on harnessing the unique properties of thorium to deliver energy solutions that are both environmentally friendly and economically viable.

The unique features of Flibe Energy's molten salt reactors set them apart from traditional reactors. These reactors use chemically stable fluoride molten salts to operate at high temperatures, leading to greater energy efficiency. And because the reactor is unpressurized, it is smaller and lighter, making it more adaptable and could enable mobility. This opens up the possibility of using these reactors as drop-in replacements for retired coal plants, potentially revolutionizing the energy industry by providing a cleaner alternative without requiring extensive infrastructure changes.[32]

Molten salt reactors' chemical reactivity and fuel efficiency are also substantial advantages. The molten salt acts as both a coolant and a solvent for nuclear fuel, allowing for continuous processing and recycling of the fuel. This means that the reactor can achieve a much higher burn-up rate, extracting more energy from each unit of fuel, and can be refueled while operating. Plus, the inherent safety features of molten salt reactors set them apart, providing a reassuring level of safety and confidence in the technology.[33]

Facing the Challenges Head On

Flibe Energy has made significant progress in advancing molten salt reactor technology. Their portfolio includes a series of pilot projects and experimental reactors, each designed to test and refine various aspects of the technology.[34] These projects include building reactors, gathering data, and learning from each iteration to improve future designs. Flibe Energy leverages a wealth of scientific expertise and technological resources in partnership with leading research institutions. These collaborations have been instrumental in overcoming the technical challenges associated with developing and deploying molten salt reactors.

Yet, despite these achievements, the path forward is not without obstacles. The development of nuclear technology is inherently complex, and molten salt reactors are no exception. Navigating the regulatory landscape is a major challenge. Nuclear technology, by its nature, requires rigorous safety standards and regulatory approval. This is a necessary safeguard, but it can also slow the pace of innovation. The challenge for Flibe, as with all innovators, is to push the boundaries of what is possible while meeting the requirements.[35]

The technical complexities of reactor design and construction also present hurdles. Building a molten salt reactor is no small feat. It requires a deep understanding of nuclear physics, chemistry, engineering, and materials science. The innovators must meticulously design and construct each component to withstand the extreme conditions within the reactor. Flibe Energy ensures that their reactors are efficient but also materially sound, secure, and compliant with international standards.[36]

Conclusion

Thorium can potentially transform how we power our civilization. In the following chapters, we'll look at how thorium can help solve major global issues like providing freshwater to drought areas, improving agriculture in dry regions, supporting clean energy for defense and disaster response, and enabling a permanent human presence in space. These aren't just theoretical ideas but are based on actual projects and emerging technologies already in progress. You'll see that thorium is not only a superior fuel but also aligns well with the future we want to create. In the next chapter, we'll start with the problem of water scarcity.

1. Alvin M. Weinberg, The First Nuclear Era (1994).
2. IAEA, The Thorium Fuel Cycle: Benefits and Challenges (2022).
3. Generation IV International Forum, "Molten Salt Reactors."
4. Oak Ridge National Laboratory, Molten Salt Reactor Experiment Safety Review (1970).
5. IAEA, Status of Molten Salt Reactor Technology (2022).
6. Oak Ridge National Laboratory, MSRE Safety Review, 18–19.
7. World Nuclear Association, "Thorium" (2023).
8. IAEA, Thorium Fuel Cycle, 10–12.
9. Oak Ridge National Laboratory, internal memo on global thorium supply, 2019.
10. International Atomic Energy Agency, The Thorium Fuel Cycle: Benefits and Challenges (Vienna: IAEA, 2022), 10–12.
11. Kirk Sorensen, TED Talk, 2011.
12. Alvin M. Weinberg, The First Nuclear Era (1994), 108-110.
13. International Energy Agency, Net-Zero Road-map (2023).
14. UN Water, World Water Development Report (2024), 15.
15. UN Water, WWDR (2024), 16.
16. International Desalination Association, "Global Desalination Market," 2023.
17. IAEA-TECDOC-1444, Introduction of Nuclear Desalination (2015).
18. IAEA, Status of Nuclear Desalination (2022).
19. US DOE Office of Nuclear Energy, "Advanced SMR Fact Sheet," 2023.
20. Australian Academy of Technology & Engineering, Nuclear Energy for Australia (2018).
21. Emirates Nuclear Energy Corporation, "Barakah Research Agenda," 2023.
22. US DOE, Coastal Resilience Options (2024).
23. International Energy Agency, Electricity 2024, 78.
24. Nuclear Innovation Alliance, Advanced Reactor Regulatory Road-map (2022).
25. Kirk Sorensen, TED Talk, 2011.
26. US Army Logistics University, "Fuel Convoy Casualty Statistics, OEF/OIF," 2014.
27. Sorensen meeting notes, author's personal files, 2013.
28. Nature commentary on Sorensen, "Thorium Reactors Revisited," 2012.

29. IEA, Energy Security in a Low-Carbon World (2021), 42.
30. Flibe Energy, "About Us" webpage, accessed 2025.
31. Flibe Energy, press kit 2024.
32. Oak Ridge National Laboratory, "Liquid Fluoride Salt SMR Concept," 2020.
33. IAEA, Near-Term Options for Thorium Deployment (2022).
34. Flibe Energy pilot-project summary, 2025.
35. Nuclear Regulatory Commission, "Advanced Reactor Licensing Modernization," 2024.
36. World Institute for Nuclear Security, Safeguards for Liquid-Fuel Reactors (2023).

Chapter 2

Water Scarcity and Thorium-Powered Desalination

A uthor's remarks:

Thinking back to when my Navy destroyer was struggling to make enough drinking water, I remember looking across the endless ocean and thinking of the Coleridge poem, "The Rime of the Ancient Mariner", as the sailors on that wooden sail ship lamented,

> *"Water, water, every where, And all the boards did shrink;*
> *Water, water, every where, Nor any drop to drink."* [1]

It made me think, with so much water in the world, why should anyone have to go thirsty?

Today, the pieces of a worldwide water solution are finally coming together. We now have the potential to power large-scale desalination and clean energy production, not with fossil fuels, but with thorium as a clean and efficient power source that promises to reshape the economy of energy and water resources.

In this chapter, we examine the problem of water scarcity, how molten-salt thorium reactors, in tandem with high-efficiency reverse osmosis (RO), could cleanly and sustainably deliver the vast volumes of freshwater that growing populations and agriculture now require, and real-world case examples of how nuclear-powered desalination can make a meaningful difference.

The Global Strain on Water Systems

In 2023, the world experienced a relentless series of climate extremes that exposed weaknesses in our global water systems. Rivers shrank, reservoirs hovered at record lows, and glaciers continued their dramatic retreat. The year closed as the hottest ever recorded. The World Meteorological Organization reported that 3.6 billion people now experience water shortages at least one month a year. That number is expected to exceed 5 billion by 2050.[2] This is no longer a distant warning. It is a crisis taking shape in real time.

Sub-Saharan Africa has become one of the most visible regions suffering from severe water stress. Subsistence farming and rain-fed crops form the backbone of daily survival. Crops that once thrived now struggle, and inconsistent rainfall and prolonged droughts have devastating effects. The impact on health is equally severe; limited access to clean water leads to widespread waterborne illnesses, so diseases spread where clean water is scarce. Governments and organizations work tirelessly to address these challenges, yet the need often overwhelms the available resources.

In the Middle East, water scarcity accelerates existing tensions. Border-crossing aquifers, rivers, and rainfall now influence political outcomes as much as agricultural ones. Access to dwindling water supplies has intensified existing instability. In Syria, years of prolonged drought between 2006 and 2010 devastated agriculture in the northeastern region, forcing over a million people to migrate to urban areas, a displacement that some analysts argue contributed to the social unrest preceding the civil war.[3] In Yemen, a nation facing long-standing economic and infrastructure challenges, water shortages in cities like Sana'a and Taiz have led to chronic public health crises and localized conflicts over wells and aquifers, further straining governance and fueling cycles of violence.[4] In contrast, water scarcity in Israel and Jordan, while severe, has become an unlikely platform for cooperation. Despite political tensions, the two nations have pursued several water-sharing agreements under the 1994 Israel-Jordan Peace Treaty. One landmark project is the Red Sea and Dead Sea Water Conveyance, a joint initiative to desalinate Red Sea water and replenish the shrinking Dead Sea, with the shared goal of addressing long-term regional water shortages. Such initiatives, while fragile, demonstrate how shared resource challenges can sometimes open channels for diplomacy where politics alone may fail.[5]

The American Southwest, particularly California and Arizona, now operates under perpetual drought alert, impacting agriculture, industry, and daily life. Once a symbol of America's ability to tame nature, the Colorado River is split thinner and thinner as more people depend on a diminishing supply. Urban centers continue expanding, increasing demands on already stressed water systems. Droughts increase the risk of wildfires, accompanied by a lack of water to fight them. The result is a complex web of challenges that require innovative solutions and collaborative efforts.

Rising Demand, Shrinking Supply: Global Pressures and the Push for Cooperative Solutions

The combined effects of climate change and population growth compound the problem. As the weather becomes less predictable, more people demand water for drinking, crops, cities, and industry. At the same time, the world's population continues to grow, with increasing urbanization exacerbating the demand for water. Cities swell with new inhabitants, stretching infrastructure to its limits and intensifying water resource competition.[6] This confluence of factors

29

makes water availability increasingly uncertain and adds to rising tensions. The result is a tightening vise.

The economic and social impacts of water scarcity are profound. Agriculture, a cornerstone of economies and societies worldwide, suffers greatly. Reduced water availability leads to lower crop yields, threatening food security and livelihoods. The ripple effect extends beyond the fields, impacting supply chains and global markets. In some regions, the scarcity of water forces families to migrate in search of better opportunities, leading to displacement and adding pressure to urban areas already struggling to cope.

In response to these challenges, global initiatives strive to address water scarcity through cooperation and innovation. The United Nations Sustainable Development Goals (SDGs), particularly Goal 6, emphasize the importance of ensuring the availability and sustainable management of water and sanitation for all.[7] This goal serves as a rallying point for international efforts, encouraging countries to invest in water infrastructure, improve water efficiency, and protect ecosystems that support water resources.

International water management agreements also play a crucial role in fostering collaboration among nations. These agreements provide frameworks for sharing water resources and resolving disputes peacefully. Promoting dialogue and cooperation helps mitigate the risk of conflict and ensure water is managed sustainably. Such initiatives are vital in a world where water knows no borders, and the need for innovative solutions and new technologies is more urgent than ever, as the challenges are shared globally. And while some regions have turned to desalination, it still requires significant energy and is often carbon intensive.

Reverse Osmosis: Water Independence at Guantanamo Bay

Turning seawater into freshwater using RO desalination (salt removal) is not new. The US naval base in Guantanamo Bay (GTMO) vividly illustrates the transformative power of reverse osmosis. In 1964, when Fidel Castro severed the power and water supply to GTMO, the US relied initially on water barges and, later, a multi-stage flash distillation plant. These methods, powered by diesel generators, were costly and inefficient, straining resources and budgets.

The early 2000s marked a turning point for GTMO as it introduced RO desalination technology. Capable of producing up to 1.4 million gallons of fresh water daily, this technology revolutionized the base's water supply.[8] The switch to RO provided a stable water source and significantly reduced energy consumption. Adding a liquefied natural gas (LNG) power plant further enhanced efficiency, achieving annual energy savings of $25 million and cutting fuel consumption by 27%.[9] This success story underscores the potential of RO to not just provide water but also reshape water management, particularly if it was instead paired with renewable energy resources, offering a repeatable approach to water sustainability.

Timeline of Converging Milestones in Thorium, Water Technology, and Climate Policy

Timeline of Converging Milestones:
Thorium, Water Technology & Climate Policy

Thorium Development | **1945**

Thorium proposed as alternative nuclear fuel

1954 | Thorium Development

Aircraft Reactor Experiment proves molten salt reactor viability

Water Technology | **1964**

Guantanamo Bay deploys desalination plant after Cuba embargo

1965 | Thorium Development

Molten Salt Reactor Experiment (MSRE) begins at Oak Ridge

Thorium Development | **1973**

U.S. defunds thorium R&D amidst Cold War uranium focus

1990 | Water Technology

Reverse osmosis becomes dominant desalination method globally

Climate Policy | **1992**

UN Framework Convention on Climate Change adopted

2006 | Thorium Development

Stern Review links climate science and economic risk

Kirk Sorensen revives interest in LFTR

2008 | Water Technology

Energy recovery systems increase efficiency of RO desalination

Thorium Development | **2011**

Flibe Energy founded to commercialize thorium MSR

2015| Climate Policy

Paris Agreement signed by 196 countries

Thorium Development | **2021**

China achieves criticality in experimental TMSR-LF1

2022 | Water Technology

Namibia launches solar-powered RO plant in Bethanie

Climate Policy | **2023**

Hottest global year on record reported by WMO

Water Desalination: How it Works

Desalination, turning saltwater into freshwater, is becoming an increasingly vital solution due to the growing global water scarcity. Several technologies exist for this, each with its strengths and trade-offs: reverse osmosis, multi-stage flash distillation, multi-effect distillation, and electrodialysis.

Reverse Osmosis Semi-Permeable Membrane Filters out Salt and Impurities

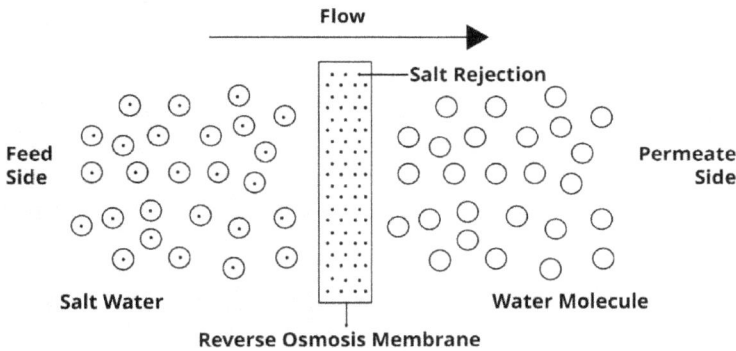

Reverse osmosis has become the most widely adopted method. In RO, seawater is forced through a semi-permeable membrane at high pressure, separating salts and impurities to yield clean, drinkable water. Although RO systems use a lot of energy, they are scalable, relatively compact, and suitable for both large and small installations. They typically require about 3-6 kWh of energy per cubic meter of water, with costs ranging from $0.50 to $2.00 per gallon.[10] Recent advancements in filter technology and energy-saving methods have made RO more cost-effective, but the high-pressure pumps still require a substantial and consistent energy source, making the pairing of desalination with reliable, low-emission energy increasingly important.

Thermal desalination methods offer a different approach

Multi-stage flash (MSF) distillation uses heat to boil seawater in successive stages, capturing the resulting vapor and condensing it into freshwater. This method is proven at large scale but requires a significant amount of energy, functioning best in countries with surplus heat or fossil fuel subsidies.

Multi-effect distillation (MED) improves upon MSF by utilizing a series of vessels at decreasing pressures, recycling heat between stages to boost efficiency. However, both methods remain dependent on abundant and cheap energy.

Electrodialysis uses electric currents to remove dissolved salts from water, making it energy-efficient and effective mainly for brackish (slightly salty) water. While it has lower energy demands ($0.5–5$ kWh/m³)[11] and is economical for treating low salinity water, it is not suitable for full-salt seawater.

In summary, the choice of desalination method often comes down to energy efficiency, operating cost, and the water source. While each technology has a role, they all require a sustainable energy source, which is where thorium-powered reactors could provide a practical advantage as the world transitions away from fossil fuels. Desalination powered by thorium represents a step toward a future where water is abundant, clean, and accessible to all.

Methods to Turn Saltwater to Freshwater

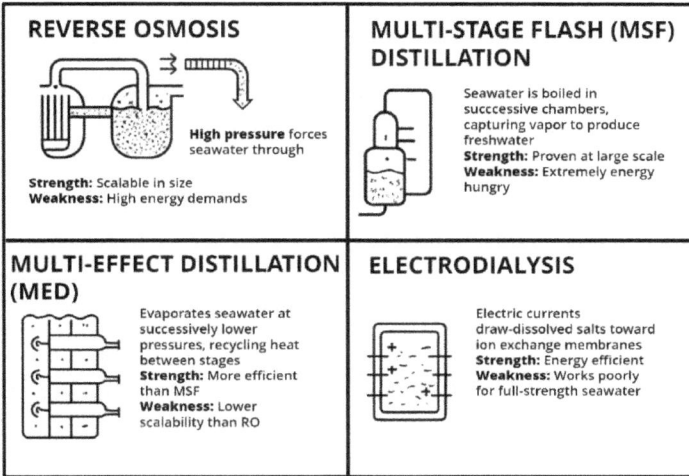

REVERSE OSMOSIS	MULTI-STAGE FLASH (MSF) DISTILLATION
High pressure forces seawater through Strength: Scalable in size Weakness: High energy demands	Seawater is boiled in succcessive chambers, capturing vapor to produce freshwater Strength: Proven at large scale Weakness: Extremely energy hungry
MULTI-EFFECT DISTILLATION (MED)	ELECTRODIALYSIS
Evaporates seawater at successively lower pressures, recycling heat between stages Strength: More efficient than MSF Weakness: Lower scalability than RO	Electric currents draw-dissolved salts toward ion exchange membranes Strength: Energy efficient Weakness: Works poorly for full-strength seawater

Desalination Innovation

Recent advancements in desalination technology provide innovations in membrane materials, such as developing more effective polymers, which have significantly improved the efficiency and durability of RO systems. Energy recovery systems, which capture and reuse energy expended during the desalination process, enhance operational efficiency, lowering costs and environmental impact. Using renewable energy sources, like solar or wind, to power desalination operations is a complementary path toward sustainable, carbon-neutral water production.

What if we could scale operations globally, especially in the driest areas where water scarcity is most acute? An investment plan involving multiple thorium reactors to create desalination hubs could meet regional energy and water needs.

Additionally, if thorium-powered desalination meets expectations, we could soon see arid regions, like parts of sub-Saharan Africa, the southwestern US, and the Middle East, turn into thriving agricultural zones with enough water for people and crops.

Brine Management: Why Thorium is a Breakthrough for Desalination

Desalination methods, especially reverse osmosis, produce brine, a concentrated saltwater waste that can disrupt marine ecosystems when dumped into the ocean. Brine may contain treatment chemicals like disinfectants and anti-scalants that pose risks to aquatic life. While brine is inevitable, it can be addressed through proper treatment. Thorium reactors offer consistent, high-output thermal and electrical energy, enabling more advanced brine management strategies.

For example:

High-pressure reverse osmosis effectively pushes seawater through membranes and produces more clean water per unit of energy. Running RO at higher pressures requires a very consistent and reliable power source, something thorium reactors can provide. Thorium's stable power supply also opens the door for hybrid desalination systems that combine RO with thermal treatment. These hybrid setups can improve the efficiency of water purification by taking advantage of the stable, around-the-clock energy and heat provided by thorium reactors.

Evaporation ponds or crystallization systems could process the brine further by using the waste heat from a thorium molten salt reactor. Instead of simply pouring concentrated brine back into the ocean, these systems would use heat that would otherwise be wasted to evaporate the remaining water, leaving behind solid salts. These salts could then be collected for industrial reuse, such as in construction materials, chemicals, or fertilizers, depending on the specific mineral content. Using waste heat in this way improves the overall efficiency of the desalination plant. It also minimizes environmental impacts by reducing marine brine discharge and creating valuable byproducts from what would have been considered waste.

Membrane distillation and **forward osmosis** are alternative desalination methods that Thorium-powered systems can also support. These methods use heat or chemical gradients, rather than just pressure, to separate water from salt. While they can be very efficient, they haven't been used widely in the past because they require a steady and controlled energy supply, something that solar or wind power doesn't consistently deliver due to weather fluctuations. A thorium reactor, on the other hand, can provide the consistent heat or electricity needed to make these systems work more reliably and with fewer added chemicals.

Zero liquid discharge (ZLD) systems can be combined with thorium-powered plants to save even more freshwater and turn left-over salts into useful products. In simple terms, ZLD means that the system squeezes out almost every drop of usable water, leaving behind solid salts that can be safely stored or sold for industrial purposes. In contrast, fossil-fueled plants are usually built for low cost and fast output, not for environmental protection, and often can't support these kinds of advanced systems.

Brine Management Strategies and Zero Liquid Discharge Feasibility

Brine Management Strategy	Process Description	Typical Energy Requirement	MSR Compatibility	ZLD Feasibility
Discharge to Ocean or Surface Water	Diluted brine is returned to sea or surface bodies; relies on sufficient mixing zones	Low	Feasible near coasts	No – Not ZLD
Evaporation Ponds	Brine is held in open ponds and water evaporates naturally or with solar assist	Very Low	Site-dependent; needs land area	Partial – Salt remains
Mechanical Vapor Compression (MVC)	Uses thermal energy and compressors to evaporate and condense water from brine	High	Well-suited with MSR heat	Yes
Multi-Effect Distillation (MED)	Sequential heat reuse to evaporate and condense water from brine	Moderate	Efficient with MSR thermal output	Yes
Crystallization	Brine is fully evaporated until solids precipitate	Very High	Enabled by high MSR temps (600°C+)	Yes – Full ZLD
Electrodialysis	Uses electric fields to move ions across membranes; best for lower salinity brines	Low to Moderate	Efficient if powered electrically	No – Not full ZLD
Salt Recovery / Mineral Harvesting	Extracts marketable salts (e.g., gypsum, lithium, magnesium) from concentrated brine	Moderate	Synergistic with MSR heat/scale	Yes (with market use)

What makes thorium-powered desalination especially promising is its ability to produce freshwater without dumping untreated waste back into fragile coastal ecosystems. In most desalination plants, brine is discharged into the ocean with little to no treatment. It is heavily composed of salts, cleaning agents, and trace metals, which can harm coral reefs, deoxygenate marine zones, and alter coastal chemistry. Thorium reactors provide consistent, carbon-free heat. This helps with advanced brine treatment methods like zero-liquid discharge, mineral recovery, and hybrid evaporation systems. These techniques turn harmful waste into useful salts and materials instead of dumping them into the ocean.

Thorium not only addresses water scarcity but also changes desalination from a necessary process into a responsible and sustainable practice. This shift from "dirty brine disposal" to "clean brine stewardship" may become a defining hallmark of next-generation desalination.

Putting it Together: A Brine Management Cost Example

In the Middle East, most desalination plants still burn oil or gas. They make fresh water, but they also pump thick, salty brine, mixed with cleaning chemicals, straight back into the sea, where it can harm coral and fish. Thorium-powered plants would work differently. Because a molten-salt reactor supplies steady, carbon-free heat and electricity all day, operators can afford extra cleanup steps that fossil-fuel plants often skip.

Let's look at the numbers. A typical Gulf-coast RO plant now spends only about 5 to 15 cents for every cubic meter of water on the simplest brine disposal, which is usually a pipe to the shoreline.[12] If the same plant adds a ZLD system that boils off almost all the water and leaves dry salt, the extra equipment pushes total water cost up by fifty to eighty cents a cubic meter.[13] When the heat comes from a thorium reactor instead of diesel, that ZLD surcharge falls to roughly twenty to thirty cents,[14] because the reactor's waste heat does much of the work.

Those extra pennies can be earned back. Dry ZLD salt is not waste. It can be sold to local chemical plants for forty to sixty dollars a ton, and high-purity salt for chlorine factories can bring close to ninety dollars.[15] A large coastal plant making 300,000 cubic meters of water a day could recover more than 100,000 tons of salt each year, about five to eight million dollars in sales. Some pilot projects in Oman and Qatar are also testing ways to harvest small but valuable amounts of magnesium and lithium from the same brine stream.[16]

Because a thorium reactor runs without interruption, it also supports hybrid designs such as reverse osmosis plus low-temperature distillation to squeeze out more fresh water with fewer chemicals. In short, pairing molten-salt reactors with desalination makes the cleanest brine solution, true ZLD, both affordable and even revenue-positive, all while cutting carbon emissions to zero.

Implementing Nuclear-powered Desalination: Case Examples

The primary hurdle is no longer scientific. It is institutional. Investment still favors known technologies, even if they are less efficient or more polluting. National policies could create an enabling environment for nuclear desalination, with clear guidelines and incentives to encourage investment and innovation.

Regulations written for uranium reactors struggle to accommodate newer designs. Regulatory oversight must uphold safety and environmental standards while allowing flexibility to adapt to new technological developments. Governments should work closely with international bodies to establish consistent regulations that support the widespread adoption of thorium technology, set standards, and facilitate cross-border collaborations.

Integrated Thorium Reactor + RO Desalination Plant

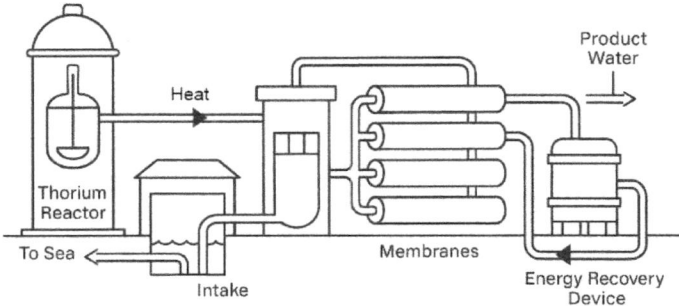

REVERSE OSMOSIS DESALINATION PLANT
POWERED BY THORIUM REACTOR

Case Examples Show Momentum Shifting Towards Nuclear Desalination

In Australia, where prolonged droughts and public resistance to uranium-based nuclear power have long shaped energy debates, thorium has emerged as a more politically friendly alternative. Researchers at the University of Adelaide and Flinders University have explored thorium molten salt reactor designs to complement large-scale RO desalination, particularly for drought-prone regions like South Australia.[17]

In the United Arab Emirates and Saudi Arabia, two nations that already rely heavily on desalination to meet municipal water needs, nuclear-powered desalination is not just theoretical. The UAE's Barakah nuclear plant began supplying grid energy that indirectly supports water treatment. Also, Saudi Arabia's King Abdullah City for Atomic and Renewable Energy (K.A.CARE) actively studies advanced small modular reactor (SMR) designs, including high-temperature gas-cooled and molten salt systems, with potential application in powering desalination plants. With both capital and climate vulnerability on their side, these countries are well-positioned to lead a transition toward thorium-powered water security.[18]

In Sub-Saharan Africa, early success stories are beginning to emerge that demonstrate how decentralized, energy-efficient desalination can transform communities. In the town of Bethanie, Namibia, a solar-powered reverse osmosis plant launched in 2022 now provides clean drinking water to the entire population, without relying on diesel or grid electricity. This off-grid solution, developed in partnership with the Namibia Water Corporation and Desert Research Foundation, underscores how scalable desalination can empower rural settlements in arid regions.[19]

Looking ahead, South Africa's Steenkampskraal project offers a glimpse of what thorium-powered water systems could achieve in the region. The site hosts one of the world's richest thorium deposits and is the proposed location for the HTMR-100, a small modular high-temperature gas-cooled reactor capable of generating both electricity and industrial heat.[20] If paired with desalination or mineral recovery, such reactors could provide clean water, synthetic fuel precursors, and grid support to communities.

Philanthropic Foundations like the Gates Foundation, the Rockefeller Foundation, and Breakthrough Energy Ventures are increasingly active in funding decentralized clean energy, health, and water infrastructure for underserved regions. While none have yet directly backed thorium-powered systems, their portfolios and mission alignments make them natural allies in demonstrating SMRs for humanitarian impact.

The Bill & Melinda Gates Foundation, through its Global Growth & Opportunity Division, has funded solar-powered water purification and health-focused energy systems in rural Sub-Saharan Africa and India. While these have focused primarily on renewables, Gates himself is also the founder and chairman of TerraPower, a nuclear innovation company pioneering advanced fast reactors, showing clear support for nuclear as a long-term clean energy solution. This positions the foundation to potentially support thorium-RO pilot systems that combine clean water with scalable nuclear reliability.[21]

Breakthrough Energy Ventures, a fund led by Bill Gates and supported by Bezos, Branson, and other tech leaders, has invested directly in advanced reactor startups such as TerraPower, Kairos Power, and Oklo. While not yet invested in thorium-specific companies, BEV's focus on high-impact, high-risk innovations that address deep decarbonization gaps positions it as a catalytic partner for thorium-RO goals. These pilots would align with BEV's climate equity goals by delivering clean water and energy to populations otherwise left behind by grid-focused investment models.

The Rockefeller Foundation has committed over $1 billion to energy access through its Global Energy Alliance for People and Planet (GEAPP), working with governments and multilateral agencies to expand distributed renewables in Africa and Asia. Though focused on solar mini-grids, the Rockefellers' climate resilience mandate and their interest in last-mile infrastructure support SMRs in geographies where sunlight is intermittent or storage remains costly. Integrating thorium SMRs into their energy-water-agriculture model could offer an emissions-free and climate-adaptive complement to renewables in high-risk regions.[22]

For climate-focused philanthropies focused on sustainable energy-water-agriculture development, thorium would provide a comprehensive solution. A thorium-powered reverse osmosis desalination project in water-scarce regions like the Sahel, Horn of Africa, or coastal South India could demonstrate how SMRs deliver reliable off-grid power for drinking water, agricultural irrigation, and local cooling needs. They offer resilience against power outages, heatwaves, and droughts. Pilot projects may provide significant humanitarian benefits and serve as proof-of-concept for scalable, low-carbon infrastructure.

Tech in Context: Where Thorium Fits in the Nuclear Innovation Landscape

As the world looks to nuclear energy to help solve climate, energy, and water challenges, multiple next-generation reactor designs are competing for attention, funding, and regulatory traction. Each aims to address the shortcomings of conventional reactors such as cost, safety, waste, and scalability, but they do so in very different ways. To understand why thorium molten salt reactors are attracting renewed interest, it's helpful to see them in context alongside other advanced technologies.

Multiple Types of Fission Reactors

Types of Nuclear Reactors

PRESSURIZED WATER REACTOR (PWR)

Uses water as coolant and moderator at high pressure, needs large containment structure.

SECONDARY STEAM

BOILING WATER REACTOR (BWR)

Water boils directly in the reactor core, simplifying the design but requiring shielding for radioactive steam.

TURBINE

CANDU REACTOR

Heavy water boosts neutron efficiency and allows natural uranium.

Heavy water boosts neutron efficiency and allows natural uranium fuel. On-power refueling enables continuous operation.

FAST BREEDER REACTOR (FBR)

Fast breeder reactor (FBR)

Uses fast neutrons without a moderator to convert fertile material into fissile fuel. Offers superior fuel efficiency.

LEAD-OR SODIUM-COOLED FAST REACTOR

Lead and sodium coolants allow for efficient fuel breeding, but require advanced materials and strict controls.

Pressurized Water Small Modular Reactors (PWR-SMRs)

Small modular reactors based on light-water technology (like those once pursued by NuScale) are smaller than conventional reactors but preserve many of their systems. They use enriched uranium fuel, operate under high pressure, and require large containment structures. Their main advantage is familiarity: regulators, utilities, and supply chains already understand them. However, they still face the same issues of spent fuel storage, water usage, and accident potential due to pressurized systems.

Sodium-Cooled Fast Reactors (SFRs)

These "fast reactors" operate with liquid sodium as a coolant and aim to recycle plutonium and leftover used fuel. They offer strong fuel efficiency and the potential to reduce long-lived radioactive waste. However, their benefits come with a unique set of engineering and safety challenges, most notably, sodium's chemical reactivity. While it transfers heat very efficiently, it reacts violently with water and burns when exposed to air, raising significant fire risks and containment concerns.

Fusion Reactors

Fusion has long been the holy grail of nuclear energy: clean, safe, and almost limitless. Projects like ITER in France and private startups around the world aim to replicate the sun's power by fusing hydrogen isotopes at extreme temperatures. While progress is exciting, commercial-scale fusion remains decades away. Critics say they defy the laws of physics. Technical hurdles like plasma containment, breeding tritium, and net energy gain are unresolved at large scales.

Thorium Molten Salt Reactors (MSRs)

Thorium MSRs represent a different category altogether. They use a liquid fuel form—thorium dissolved in molten fluoride salts—which circulates at atmospheric pressure. This eliminates the risk of pressure-related explosions and allows for passive safety features like freeze plugs and gravity-driven shutdowns. Thorium itself cannot be used directly in weapons, and its fuel cycle produces less long-lived waste. The reactor can operate at high temperatures, making it ideal for both electricity and process heat applications like desalination. What sets thorium MSRs apart is not just that they're safer, but that they're fundamentally more efficient, adaptable, and environmentally aligned.

Why Thorium Stands Out

While fusion is still years away and PWR-SMRs and sodium-cooled reactors optimize existing paradigms, thorium MSRs offer something rarer: a truly different model. One that's passive in safety, compact in form, abundant in fuel, and work well with other systems like desalination, industrial heat, synthetic fuel production, or off-grid deployment. Thorium doesn't just work better in theory. It matches the problems we face today.

Summary: A Future of Clean Water, Clean Energy

Imagine this: a coastline near you dotted with compact thorium-RO hubs. They hum quietly, drawing seawater through pipelines, filtering out salt, and delivering fresh water to farms and cities. They run day and night, unaffected by weather or imported fuel prices. Their footprint is small. Their carbon emissions are zero. Their waste heat is reused, and the brine is cleanly managed.

In this chapter, we've looked at case examples worldwide that could lead to such a future. The lessons learned from these projects are repeatable successes as we consider future expansion opportunities. The scalability of thorium-powered desalination offers the potential for replication in similar regions worldwide. Innovation continues to drive improvements in desalination technologies, making them more efficient and cost-effective. Countries facing severe water shortages can see these case studies as blueprints for success, adapting the strategies and technologies used to suit their unique needs and circumstances.

So why was this technology overlooked for so long? In the next chapter, we'll take a deeper look at how this tech works, how it fell by the wayside, and the new resurgence of thorium-fueled innovation.

1. SAMUEL TAYLOR COLERIDGE, THE RIME OF THE ANCIENT MARINER (1834 TEXT).

2. WORLD METEOROLOGICAL ORGANIZATION, STATE OF GLOBAL WATER RESOURCES 2023 (GENEVA, 2024).

3. FRANCESCO FEMIA AND CAITLIN WERRELL, "SYRIA: CLIMATE CHANGE, DROUGHT, AND SOCIAL UNREST," CENTER FOR CLIMATE AND SECURITY (2012); PETER H. GLEICK, "WATER, DROUGHT, CLIMATE CHANGE, AND CONFLICT IN SYRIA," WEATHER, CLIMATE, AND SOCIETY 6 (2014): 331–340.

4. PETER H. GLEICK, IBID.; ADDITIONAL PUBLIC-HEALTH DATA IN YEMEN MINISTRY OF WATER AND ENVIRONMENT, NATIONAL WATER SECTOR STRATEGY (2021).

5. WORLD BANK, RED SEA–DEAD SEA WATER CONVEYANCE STUDY PROGRAM: ENVIRONMENTAL AND SOCIAL ASSESSMENT – EXECUTIVE SUMMARY (WASHINGTON, DC, 2014); ISRAEL–JORDAN PEACE TREATY, ANNEX II (1994).

6. US GEOLOGICAL SURVEY, COLORADO RIVER BASIN WATER SUPPLY AND DEMAND STUDY: 2022 UPDATE TO CONGRESS (CIRCULAR 1491, 2022).

7. UNITED NATIONS GENERAL ASSEMBLY, TRANSFORMING OUR WORLD: THE 2030 AGENDA FOR SUSTAINABLE DEVELOPMENT (A/RES/70/1, 2015), GOAL 6.

8. AMERICAN MEMBRANE TECHNOLOGY ASSOCIATION, "GUANTANAMO BAY SEAWATER RO DESALINATION PLANT ENERGY RECOVERY SYSTEMS," CASE STUDY, ACCESSED 2025.

9. ENERGY RECOVERY INC., "OPTIMIZING REVERSE OSMOSIS PERFORMANCE THROUGH THERMAL INTEGRATION," TECHNICAL PAPER 2022.

10. INTERNATIONAL ATOMIC ENERGY AGENCY, INTRODUCTION OF NUCLEAR DESALINATION: A GUIDEBOOK (IAEA-TECDOC-1444, 2015), 112–15.

11. INTERNATIONAL ATOMIC ENERGY AGENCY, STATUS OF NUCLEAR DESALINATION (VIENNA, 2022), 57–60.

12. GLOBAL WATER INTELLIGENCE, DESALINATION MARKETS 2024, EXECUTIVE DATASET (OXFORD, 2024), TABLE 8.

13. AMIRHOSSEIN RAHBAR, "COST ANALYSIS OF ZERO-LIQUID-DISCHARGE DESALINATION IN THE GULF," JOURNAL OF WATER PROCESS ENGINEERING 54 (2023): 102617.

14. INTERNATIONAL ATOMIC ENERGY AGENCY, ECONOMIC ASSESSMENT OF NUCLEAR-POWERED DESALINATION (VIENNA, 2023), 32–35.

15. GULF CHLOR-ALKALI PRODUCERS ASSOCIATION, "MARKET PRICES FOR SOLAR-GRADE SALT," PRICE BULLETIN Q4 2024.

16. QATAR ENVIRONMENT & ENERGY RESEARCH INSTITUTE, PILOT-SCALE MINERAL RECOVERY FROM DESALINATION BRINE, project brief, 2024.

17. FLINDERS UNIVERSITY, "EXPERTS EXPLORE THORIUM REACTORS FOR SAFER, CLEANER ENERGY FUTURE," news release, 27 OCTOBER 2020.

18. EMIRATES NUCLEAR ENERGY CORPORATION, BARAKAH NUCLEAR ENERGY PLANT FACT SHEET (ABU DHABI, 2023); K.A.CARE, "SMALL MODULAR REACTOR ROAD-MAP FOR DESALINATION," BRIEFING PAPER 2024.

19. NAMIBIA WATER CORPORATION LTD. AND DESERT RESEARCH FOUNDATION OF NAMIBIA, "BETHANIE HYBRID SOLAR-POWERED DESALINATION PLANT COMMISSIONED," project summary, JULY 2022. HTTPS://WWW.DRFNAMIBIA.ORG.NA/BETHANIE-DESALINATION

20. STEENKAMPSKRAAL THORIUM LIMITED, "HTMR-100: HIGH TEMPERATURE MODULAR REACTOR OVERVIEW," company white paper, ACCESSED JUNE 2025. HTTPS://WWW.THORIUM100.COM/

21. TERRAPOWER. "OUR MISSION." TERRAPOWER.COM, ACCESSED JUNE 2025. HTTPS://WWW.TERRAPOWER.COM/ABOUT

22. THE ROCKEFELLER FOUNDATION. "GLOBAL ENERGY ALLIANCE FOR PEOPLE AND PLANET." ACCESSED JUNE 2025. HTTPS://WWW.ROCKEFELLER-FOUNDATION.ORG/GEAPP

Part Two

Part II: Innovation Ignited

Chapter 3

Innovation and the Thorium Breakthrough

*A*uthor's remarks:

My conviction about thorium deepened when I discovered the story of Alvin Weinberg.[1] He was a physicist and a systems thinker in an era of Cold War urgency and the race for weapons dominance. While others focused on refining uranium and plutonium for bombs and reactors, Weinberg envisioned something else entirely: a reactor that could be safer, smaller, and simpler. A reactor that used thorium.

As the director of Oak Ridge National Laboratory, Weinberg led the Molten Salt Reactor Experiment in the 1960s. His aim wasn't just efficiency. It was safety. He understood that a reactor must be robust and inherently safe, a principle that resonated with me during my time on nuclear submarines. His journal, *Nuclear Safety*, underscored the importance of safety in nuclear technology. He proposed a liquid-fueled thorium reactor cooled by molten salts, operating at low pressure and capable of shutting down safely without human intervention.[2] Far safer, this type of reactor does not use a high-pressure steam system like the reactor designs that later had meltdowns at Chernobyl and Fukushima.[3]

Unfortunately, Weinberg's ideas ran counter to the prevailing military-industrial momentum.[4] Uranium reactors supported weapons programs. Thorium reactors did not. These elements had the backing of powerful entities within the Atomic Energy Commission (AEC), making it difficult for alternative technologies like Weinberg's molten salt reactor to gain traction. Additionally, securing funding for his projects proved difficult, as the focus remained on the more established nuclear technologies. Weinberg faced skepticism and resistance in his advocacy for thorium, but his prototypes demonstrated the viability of the technology. Today, we are rediscovering his legacy.[5]

What Makes the Thorium Cycle Different

Unlike traditional nuclear reactors that use solid fuel pellets, a thorium molten salt reactor (MSR) uses fuel in liquid form. The fuel is a hot, flowing mixture of lithium fluoride and beryllium fluoride salts, with thorium and uranium dissolved in it. This liquid flows through a reactor core, where neutrons from fission reactions strike thorium-232 atoms, ultimately turning them into uranium-233. That isotope then undergoes fission, releasing energy. This transformation is important because thorium is a more abundant fuel but is not directly fissile, meaning it cannot sustain a chain reaction until it becomes uranium-233, which can then repeat the cycle via a nuclear chain reaction.

What causes the transformation? The chain reaction can be briefly summarized. When thorium-232 is placed in the reactor and absorbs a neutron, it becomes thorium-233. It is unstable and undergoes beta decay into protactinium-233, which decays over a few days into uranium-233. When a neutron hits uranium-233, it splits (fissions), releasing energy and more neutrons, which can then convert more thorium into uranium, continuing the cycle.[6]

The Thorium Fuel Cycle in a Molten Salt Reactor

THE THORIUM FUEL CYCLE

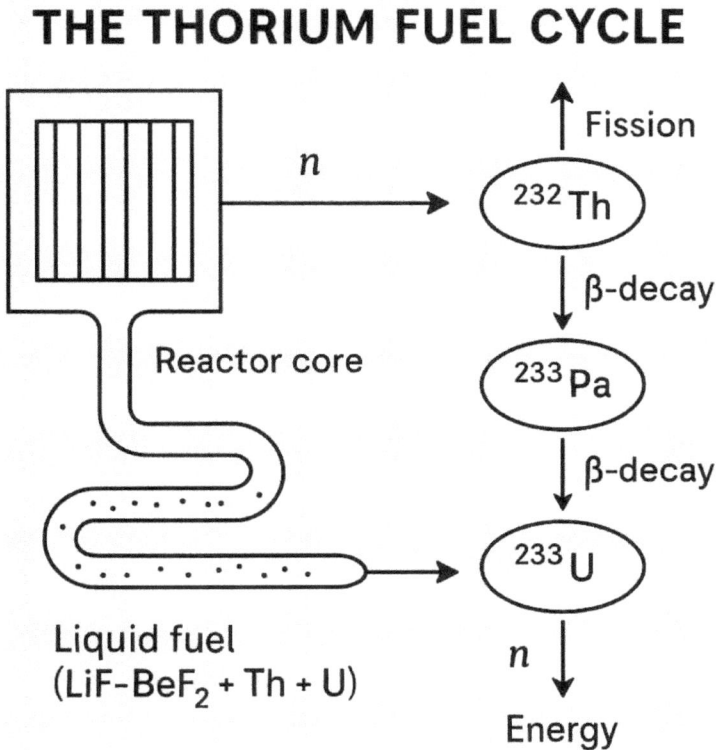

Fission

n

^{232}Th

β-decay

Reactor core

^{233}Pa

β-decay

^{233}U

Liquid fuel
$(LiF-BeF_2 + Th + U)$

n

Energy

This cycle creates very little long-lived waste. It also includes the freeze plug built-in safety mechanism as described in Chapter 1.[7] There's no pressure build-up. No runaway heat. No meltdown. Additionally, thorium is far more abundant than uranium,[8] and it becomes clear why so many researchers and engineers are circling back to Weinberg's ideas.

Energy Efficiency and Safety

Thorium reactors extract more energy per unit of fuel than conventional designs.[9] They also operate at higher temperatures and lower pressures, particularly when using molten salt as a coolant. This makes them more efficient for generating electricity and less dependent on massive water-cooling systems.

Thorium is unmatched among nuclear and fossil fuel sources regarding energy efficiency and output. Most nuclear reactors today use solid uranium fuel, but less than one percent of that fuel actually gets used. The rest becomes radioactive waste because the design doesn't allow the fuel to be reused. Thorium reactors use a different method. They work with liquid fuel that can be cleaned and reused through reprocessing.

In studies where efficiency is based on continuous reprocessing (not yet proven in commercial reactors), thorium is expected to yield a higher energy output per unit, estimated to be 200 times more efficient than uranium. In other words, for every ton of thorium fuel, we'd need 200 tons of uranium fuel.[10]

Now, what does that mean for us? It means the reactor can filter out the leftovers, keep the useful parts flowing, and squeeze far more energy out of the same material. This not only boosts efficiency, but also cuts down on the amount of waste, and how dangerous that waste is over time.

When reprocessing is built into the system, nearly all the thorium can be turned into usable fuel. That fuel then splits and releases energy (fission), and the process repeats. The waste that's left behind? Instead of remaining radioactive for tens of thousands of years, as in today's uranium reactors, it drops to safe levels in just a few hundred years.[11] That's a big shift.

But it only works if reprocessing is done right. It must occur continuously during operation to maintain the fuel's cleanliness and prevent the accumulation of unwanted isotopes. When that's in place, thorium reactors produce energy efficiently and leave behind far less long-term radioactive material. It's a smarter, cleaner nuclear future.

Fission Product Half-Lives: Thorium vs. Uranium

Category	Thorium Fuel Cycle	Uranium Fuel Cycle
Dominant Long-Lived Isotope	Uranium-233 (fissile); Fission products decay within decades	Plutonium-239 and minor actinides with half-lives >10,000 yrs
Fission Products > 30 years	Very few; decay products drop below background in ~300 yrs	Many, including Technetium-99 and Iodine-129
Residual Heat After 1000 Years	Negligible	Significant from minor actinides and long-lived isotopes
Radiotoxicity After 500 Years	Comparable to natural uranium	Still high—requires deep geological storage
Disposal Strategy	Near-surface or intermediate-depth disposal possible	Requires long-term geological isolation
Proliferation Risk of Byproducts	Low (U-232 contamination complicates weapons use)	High (Pu-239 separable in spent fuel)
Transuranic Waste Generation	Minimal	Substantial

One of the big advantages of molten salt reactors is that the fuel is a liquid. That might not sound like a big deal at first, but it opens the door to something really important: cleaning the fuel while the reactor is still running. As the fuel circulates, it builds up fission products: the leftovers from splitting atoms. In a molten salt system, those can be gradually removed using new techniques like distillation or chemical separation. This helps prevent the buildup of unwanted materials, which can slow down the reaction or shorten fuel life.

These cleanup methods are still being developed and tested, but the core idea is proven. A working molten salt reactor can be designed to clean itself periodically, keeping the fuel efficient and extending the time between refueling.

Some critics say this just swaps one kind of nuclear waste for another. But even including the waste from fission products, the total volume of long-term radioactive material is far smaller than in today's uranium reactors. And here's another key point: it doesn't need the same kind of deep, long-term storage. What's left is lower in volume, less hazardous over time, and can often be stored in simpler facilities for a few hundred years instead of tens of thousands.[12]

From a safety perspective, thorium reactors offer multiple layers of protection. They rely on passive systems, which means they don't need a human to hit a panic button if something goes wrong. Unlike uranium reactors, thorium doesn't produce plutonium as a byproduct, so it's much harder to weaponize.[13] For people worried about nuclear accidents, this isn't just a better design; it's a different equation entirely. Instead of trying to manage every way a meltdown could happen, thorium's design helps prevent those paths from even forming.

Why Thorium Was Overlooked

In the early days of nuclear energy, the world had a choice. Uranium reactors were chosen not because they were best for civilian energy but because they were dual-purpose. They produced electricity and plutonium.[14]

Thorium didn't serve that military role. So, it was shelved. Research continued in small pockets—at Oak Ridge, in India, and in parts of Europe—but without the funding or urgency of uranium programs.

Timeline: Why Thorium Was Passed Over

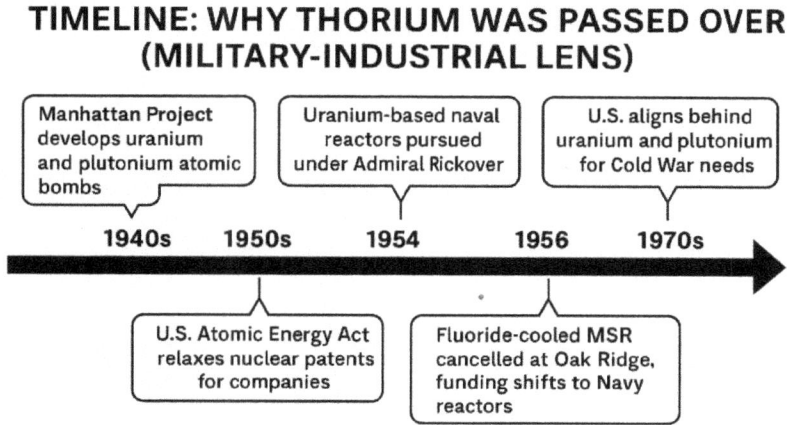

TIMELINE: WHY THORIUM WAS PASSED OVER (MILITARY-INDUSTRIAL LENS)

Manhattan Project develops uranium and plutonium atomic bombs	Uranium-based naval reactors pursued under Admiral Rickover	U.S. aligns behind uranium and plutonium for Cold War needs

1940s   1950s   1954   1956   1970s

U.S. Atomic Energy Act relaxes nuclear patents for companies	Fluoride-cooled MSR cancelled at Oak Ridge, funding shifts to Navy reactors

That history matters because it means thorium wasn't a technological failure. It was a political casualty.[15]

In the 21st century, with the growing need for clean energy and a reduced appetite for nuclear weapons, thorium is getting a second look. And not just from governments.

The Thorium Cycle: Smaller, Safer, More Efficient Reactors

Imagine a reactor small enough to fit on an airplane, eliminating the need for the towering cooling stacks and massive infrastructure typical of traditional nuclear power plants. The thorium fuel cycle used in a molten salt reactor is so small and efficient that it was studied for potential application in aircraft.[16] Why not build nuclear-powered bomber aircraft if nuclear-powered submarines could stay on constant patrol? In 1954, the US Air Force envisioned nuclear-powered bombers that could remain airborne for weeks or even months, providing virtually unlimited range to serve as a continuous nuclear deterrent. A land-based Aircraft Reactor Experiment studied a molten salt thorium reactor for this use.

So why aren't there reactor-powered aircraft today? Because it's not practical for aircraft. The main challenges were heat (the reactor operates at 600 to 900 degrees Celsius), which increases weight from using heavier parts, plus complexity and safety concerns. In-flight refueling and ballistic missiles eliminated the need for aircraft reactors. Nevertheless, the Aircraft Reactor Experiment successfully operated for about 100 hours.[17] It proved the viability of molten salt reactors, which inspired Alvin Weinberg to conduct his Molten Salt Reactor Experiment in the 1960s, paving the way for thorium molten salt reactors today.

Summary: What Still Needs to Happen – From Challenge to Momentum

Thorium molten salt reactors face several technical challenges, but these can be solved relatively soon. The molten salts used as fuel and coolant are chemically aggressive, which means we need special materials that can resist corrosion and high temperatures of around 600 degrees Celsius. Certain alloys have shown potential in tests[18] and are being re-evaluated with modern metallurgy and testing capabilities. However, scaling these materials for commercial use remains tricky and expensive. Additionally, the thorium fuel cycle complicates the management of uranium-233,[19] especially due to contamination from uranium-232, which emits hard gamma rays. This raises some technical and slight safety concerns, but they can be managed with careful reactor design.

Beyond laboratory corrosion tests, these materials must be produced in ton-scale melts, forged or cast into vessels and pump housings, and then qualified under nuclear pressure-boundary codes. Hastelloy-N, one of the alloys, depends on a nickel content above sixty-five percent, and global aerospace-grade nickel supply is already tight. Specialty mills now turn out only a few thousand tons per year, most under long-term turbine-blade contracts.[20]

Building even a handful of commercial MSRs would therefore require specialized equipment and materials that aren't widely available yet. This includes new types of vacuum furnaces, special welding materials that can handle hot fluoride salts, and updated safety and design standards for these high-temperature reactors.[21] On the materials side, one key ingredient—lithium-7—needs to be separated from lithium-6, and only one plant in the world, in China, produces the ultra-pure lithium needed for reactors.[22]

The supply of beryllium-fluoride capacity is also limited, since it's expensive to safely handle the toxic beryllium dust, and only a few manufacturers have invested in the necessary equipment.[23]

Until these supply bottlenecks are resolved, MSR vendors will struggle with firm capital estimates or delivery schedules, even if their core reactor designs are sound. Managing the thorium fuel cycle is complex, particularly in isolating uranium-233 and mitigating contamination from uranium-232. This challenge raises technical and slight safety concerns, but international research institutions are collaborating on safer, more precise fuel handling methods.

Beyond technical hurdles, developing small modular reactors requires substantial investment [24] from motivated investors or governments. A significant obstacle is the outdated regulatory framework,[25] which is designed for traditional solid-fuel light-water reactors. The current licensing processes often do not consider the unique aspects of MSRs. Without updated regulations and pilot projects, thorium MSRs may remain experimental despite the science being understood.

Public perception of nuclear energy has become more favorable, aided by organizations like The Thorium Energy Alliance,[26] which informs policymakers and the public about the benefits of thorium. Countries like India and China actively build thorium reactor prototypes[27], while US national labs support next-generation reactor designs. Investors are beginning to recognize the long-term potential of low-waste, high-efficiency nuclear systems. If governments can modernize regulations and if innovators can build trust with communities, thorium could redefine what nuclear means in our time.

Today's ecosystem includes private-sector firms in the US, Europe, and Asia advancing commercial thorium reactor designs, supported by government research funding and updated licensing frameworks. From national labs in Canada to demonstration reactors in China, the effort has moved beyond theory into a global race to refine, test, and prove this technology.

Rather than being stalled by complexity, today's innovators are treating these challenges as solvable engineering problems. Thorium's story is no longer about "if," but "how soon." From reactors the size of trailers to AI-managed chemistry, thorium's future is modular, mobile, and transformative. But what are the options for designing a reactor? What is the outlook for small modular reactors? The next chapter explores next-generation designs.

1. Alvin M. Weinberg, The First Nuclear Era: The Life and Times of a Technological Fixer (1994).
2. OAK RIDGE NATIONAL LABORATORY, MOLTEN SALT REACTOR EXPERIMENT FINAL REPORT (1972).
3. INTERNATIONAL ATOMIC ENERGY AGENCY (IAEA), CHERNOBYL ACCIDENT SUMMARY REPORT (1986); IAEA, FUKUSHIMA DAIICHI ACCIDENT REPORT (2015).
4. US ATOMIC ENERGY COMMISSION, HISTORY AND TECHNICAL PROGRESS OF NUCLEAR POWER (1968).
5. HARGRAVES, ROBERT. THORIUM: ENERGY CHEAPER THAN COAL (WHITE RIVER JUNCTION: CHELSEA GREEN PUBLISHING, 2012), PREFACE, 168–170.
6. IAEA, THE THORIUM FUEL CYCLE: BENEFITS AND CHALLENGES (2022), 4–7.
7. OAK RIDGE NATIONAL LABORATORY, MSRE SAFETY REVIEW (1970), 12–15.
8. WORLD NUCLEAR ASSOCIATION, "THORIUM," INFORMATION PAPER, UPDATED 2023.
9. GENERATION IV INTERNATIONAL FORUM, "MOLTEN SALT REACTORS," TECHNOLOGY STATUS PAPER, 2021.
10. IAEA, NEAR-TERM AND PROMISING LONG-TERM OPTIONS FOR THORIUM DEPLOYMENT (2022), 9.
11. Oak Ridge National Laboratory, MSR Fuel Utilization Briefing, internal research memo (2010); Kirk Sorensen, TED Talk, 2011; Robert Hargraves, New Nuclear is Hot (Hanover, NH: createspace, 2025), chap. 3.
12. IS THORIUM THE FUTURE OF NUCLEAR POWER?", YOUTUBE VIDEO, 13:49, KYLE HILL, POSTED NOVEMBER 29, 2023, HTTPS://WWW.YOUTUBE.COM/WATCH?V=7AZ1DJHW7XW.
13. Generation IV International Forum, above, 13.
14. Spencer R. Weart, Nuclear Fear: A History of Images (1988), 189.
15. Weinberg, First Nuclear Era, 243–45.
16. Oak Ridge National Laboratory, AIRCRAFT REACTOR EXPERIMENT TECHNICAL REPORT (1955).
17. Ibid., 22–24.
18. Oak Ridge National Laboratory, "Corrosion Performance of Hastelloy-N in MOLTEN FLUORIDE SALTS," MATERIALS REPORT, 2020.

19. IAEA, Management of Uranium-233 (Technical Report Series #481, 2018).

20. International Nickel Study Group, Nickel Market Outlook 2024 (Lisbon, 2024), 11–13.

21. ASME Boiler and Pressure Vessel Code, Section III, Division 5, "High-Temperature Reactors," working draft (New York, 2023).

22. World Nuclear Association, "Lithium-7 Supply for MOLTEN SALT REACTORS," TECHNICAL BRIEF, UPDATED 2023.

23. US OCCUPATIONAL SAFETY AND HEALTH ADMINISTRATION, BERYLLIUM IN MANUFACTURING: COMPLIANCE GUIDE (WASHINGTON, DC, 2022), 4–6.

24. International Energy Agency, Nuclear Power in Clean-Energy Transitions (2023), 54–56.

25. Nuclear Innovation Alliance, Advanced Reactor Licensing ROAD-MAP (2022).

26. Thorium Energy Alliance, "Policy Brief: Thorium for Secure Energy and Water," 2024.

27. World Nuclear News, "China Starts Up First Thorium MSR Pilot," 2021; Department of Atomic Energy (India), "AHWR-LEU-300 Status Update," 2024.

Chapter 4

The Reactor Blueprint: Comparing Technologies for a Clean Energy Era

I n every major shift in how we produce energy, the materials, designs, and decisions made at the foundational stage have determined the course of history. Today's nuclear innovators face the same critical choices. Each reactor design, whether it's based on uranium, thorium and molten salts, or fast neutrons, comes with trade-offs in safety, efficiency, sustainability, and scalability.

The choices we make now will define how quickly, safely, and affordably we can build the clean energy systems the world demands. In this chapter, we look deeply inside the engineering decisions behind fuels, coolants, moderators, and reactor architectures. We also examine the rise of small modular reactors (SMRs) and the potential of molten salt systems. These choices are not purely academic; they could make a difference in the successful performance of rural micro-grids, disaster recovery efforts, and the production of clean water and synthetic fuel.

Reactor Fuels: What We Burn for Power

At the heart of any nuclear reactor is its fuel, the material that splits apart, or undergoes fission, to release energy. For decades, the main fuels have been uranium-235 and plutonium-239.

Uranium-235, used in most commercial reactors, generates a lot of energy but also produces long-lasting radioactive waste. Highly enriched uranium (HEU), with over 20% U-235 content, powers naval vessels and research reactors due to its compact energy density. However, HEU also poses serious risks, as it can be used to make nuclear weapons.[1]

Plutonium-239, which is usually created from uranium-238, is used in breeder reactors. While efficient, it's also a weapons-grade material, making its handling and storage highly sensitive and politically controversial.

Thorium-232, in contrast, is more abundant in nature and cannot be directly weaponized, making it an attractive option for sustainable nuclear energy. Thorium, distinguished by its silvery hue and named in honor of Thor, the Norse deity of thunder, is three to four times more naturally occurring than uranium, primarily found in igneous rock and heavy mineral sands. While thorium can't directly fuel a nuclear reaction, it can be converted into uranium-233, a material that is fissile and can sustain nuclear reactions. The fuel cycle of thorium produces less long-lived waste and offers inherent safety advantages.[2]

Mixed oxide (MOX) fuel, combining uranium dioxide and plutonium dioxide, can recycle spent fuel, but handling the plutonium content raises radiation and security concerns.

Metallic fuels, used in some fast reactors, provide excellent heat conduction but suffer from swelling and reactivity risks under intense radiation.

Reactor Fuel Comparison

Fuel Type	Composition	Common Uses	Advantages	Disadvantages
Uranium-235	Enriched uranium	Traditional reactors	High energy yield	Long-lived waste, proliferation risk
Plutonium-239	Reprocessed uranium	Breeder reactors	Reusable fuel cycle	Security concerns, complex processing
Thorium-232	Natural thorium	Advanced reactors	Abundant, produces less waste	Requires conversion to uranium-233
MOX Fuel	Uranium dioxide + plutonium dioxide	Recycled reactor fuel	Enables fuel recycling	Radiation handling risks, plutonium proliferation concerns
Metallic Fuels	Uranium or plutonium metal alloys	Fast reactors	Excellent heat conduction	Swelling, reactivity under radiation, structural stability risk

Coolants: Keeping the Core Stable

The coolant is an equally important choice in reactor design. Coolants carry heat away from the reactor core and help convert thermal energy into electricity.

Water is the most common coolant, especially light water (ordinary H_2O). It's practical and familiar but operates under high pressure, requiring heavy infrastructure and enriched fuel. Light-water reactors typically function at pressures around 150 atmospheres (over 2,000 psi) to prevent boiling, which requires robust containment structures and complex safety systems. While widely adopted due to decades of operational experience, this high-pressure environment also increases construction and maintenance costs and limits thermal efficiency compared to high-temperature, low-pressure alternatives like molten salt.

Heavy water (D_2O), used in Canada's CANDU reactors, absorbs fewer neutrons than regular water, leaving more neutrons in play to sustain a chain reaction using natural, unenriched uranium.[3] However, it's expensive to produce. The global supply of heavy water is concentrated in a few countries, primarily Canada, India, Argentina, and historically Norway, with Canada and India being the most significant current producers. Because heavy water is considered a dual-use material, trade is tightly regulated under international non-proliferation agreements, and exports can be delayed or restricted due to geopolitical factors. Limited production capacity, complex transport logistics, and dependence on a few suppliers pose additional challenges for countries without domestic production capabilities.

Liquid sodium, used in fast breeder reactors, offers excellent thermal conductivity at low pressure but reacts violently, even explosively, with air or water, making safety challenging. Furthermore, its opacity can hinder inspection.

Molten salts, such as FLiBe (lithium fluoride and beryllium fluoride), operate at normal atmospheric pressure and high temperatures, improving thermal efficiency. They're also chemically stable, but their corrosive nature requires advanced materials. Molten salt's ability to dissolve fuel directly adds design flexibility but demands precise chemical control.[4]

Lead, or Lead-Bismuth Eutectic (LBE) coolants, are used in some advanced nuclear reactor designs, particularly fast reactors like Lead-cooled Fast Reactors (LFRs). These heavy metal coolants are attractive for several reasons. First, they are corrosion-resistant when paired with suitable structural materials, offering long-term durability at high temperatures. Second, they are "neutron-efficient" because they do not absorb many neutrons. This helps the reactor use neutrons more effectively, which improves its ability to produce new fuel or break down long-lived nuclear waste. However, lead and lead-bismuth are extremely dense, which adds considerable weight to reactor systems and places structural demands on the reactor vessel and support components. LBE introduces additional challenges. While lead itself is relatively benign, bismuth can transmute under neutron bombardment into polonium-210, a highly radioactive and toxic isotope, requiring strict containment protocols. Despite these complexities, LFRs are researched for their potential to deliver high-temperature heat, long core lifetimes, and passive safety.

Helium, by contrast, is a noble gas used in High-Temperature Gas-cooled Reactors (HTGRs) such as the Pebble Bed Reactor (PBR) and Prismatic Block Reactor designs. Helium is chemically inert, meaning it does not react with fuel, structural materials, or moderators, even at extreme temperatures. It is also thermally stable and remains in a gaseous state over a wide range of temperatures and pressures. This makes it an excellent choice for reactors operating above 700°C. These high operating temperatures enable efficient electricity generation and offer possibilities for industrial applications such as hydrogen production. The main challenge with helium is its low density, which limits its ability to carry heat efficiently. To overcome this, helium-cooled reactors must operate at very high pressures, typically up to 7 MPa (about 1,000 psi), to achieve the necessary heat transfer. This demands robust, leak-tight pressure vessels and advanced gas handling systems, as helium atoms are tiny and prone to leakage. Despite these engineering hurdles, helium remains one of the most promising coolants for next-generation reactors aimed at maximizing efficiency and minimizing chemical risks.

Carbon dioxide is used as a coolant in Advanced Gas-cooled Reactors (AGRs), a nuclear reactor developed and operated primarily in the United Kingdom. AGRs are a second-generation reactor design that evolved from the earlier Magnox reactors, using graphite as a neutron moderator and carbon dioxide gas as the primary coolant. CO_2 is chosen because it is chemically stable under reactor conditions, does not become radioactive, and easier to handle than liquid metal coolants. CO_2 performs well at high temperatures, allowing AGRs to reach outlet temperatures of around 650°C, significantly higher than typical light-water reactors. This results in greater efficiency for electricity generation and supports industrial applications that require high-temperature heat. However, CO_2 doesn't transfer heat as efficiently as water, so AGRs require large heat exchangers and high coolant flow rates to maintain performance. Despite these engineering trade-offs, using carbon dioxide in AGRs represents a practical balance between safety, thermal performance, and operational maturity.

Reactor Coolant Comparisons

Coolant Type	Common Reactors	Advantages	Disadvantages
Water	PWR, BWR	Abundant, familiar	High pressure, potential for leaks
Sodium	Fast Breeder Reactors	High thermal conductivity, low pressure	Reactive with water, complex handling
Molten Salt	Molten Salt Reactors	Atmospheric pressure, high efficiency	Requires corrosion-resistant materials
Lead / LBE	Fast Reactors, SMRs	Chemically inert, high boiling point, good neutron economy	Heavy, causes structural stress, material corrosion at high temps
Helium	High-Temperature Gas Reactors	Chemically inert, low neutron absorption, no phase change	Requires high pressure, low heat capacity
Carbon Dioxide	Gas-Cooled Reactors	Inert, stable, easy to handle	Less effective heat transfer, operates at high pressure

Moderators: Slowing Neutrons for a Chain Reaction

Moderators slow down fast-moving neutrons, increasing the likelihood of fission. The choice of moderator impacts operational costs and risks such as fire hazards.

Light water is just regular water and is the most common material used to cool and slow down neutrons in many nuclear reactors. However, it has a drawback: it absorbs some of the neutrons required to sustain the chain reaction. To compensate for this, the uranium fuel must be enriched. Enrichment means increasing the amount of uranium-235, the part of natural uranium that actually splits and releases energy. Since natural uranium contains only a tiny amount of U-235, boosting its concentration is necessary for the reactor to operate efficiently with light water.

Heavy water, valued for its low neutron absorption, allows reactors to use natural (unenriched) uranium as fuel, a considerable advantage in fuel processing. However, producing heavy water is a technically complex and expensive process. It requires the separation of deuterium, a heavier isotope of hydrogen, from ordinary water, typically through energy-intensive processes like distillation or electrolysis. These methods require specialized infrastructure and yield relatively low amounts, making heavy water both costly and limited in availability. The global supply is concentrated in a few countries, adding logistical and geopolitical hurdles to its widespread adoption.

Graphite is a material used in some nuclear reactors, such as RBMK and Magnox, as a moderator, which means it helps slow down neutrons so the reactor can operate efficiently. It works well at high temperatures, but it has a downside: if it comes into contact with air or water while hot, it can ignite, posing a serious safety risk. That's part of what made the Chernobyl disaster so dangerous. Graphite is bulky and costly, but it's great at slowing down neutrons without absorbing them. This makes it especially useful in reactors that use natural or only slightly enriched uranium fuel, as it helps sustain the chain reaction without requiring much extra fuel.

Beryllium and beryllium oxide are materials that help slow down neutrons very efficiently, which is important in nuclear reactors. They're also great at handling heat. Because of these qualities, they're useful in advanced and compact reactor designs where space is tight and heat needs to be managed well. However, beryllium is expensive because it is rare, difficult to extract, and costly to refine. It is also highly toxic; inhalation of beryllium dust or fumes can cause a chronic and potentially fatal lung condition known as berylliosis. This combination of health hazards and high production costs limits its widespread use, particularly in commercial-scale reactors.

Reactor Moderator Comparisons

Moderator Type	Advantages	Disadvantages	Common Use
Light Water	Widely available, well-understood technology	Absorbs neutrons, requires enriched uranium	PWRs, BWRs
Heavy Water	Low neutron absorption allows use of natural uranium	Expensive and complex to produce, limited global supply	CANDU reactors
Graphite	Excellent neutron moderation, tolerates high temperatures	Fire hazard if exposed to air, bulky and costly, can oxidize or ignite	RBMK, Magnox reactors
Beryllium	High neutron efficiency, excellent for compact designs	Toxic, expensive, difficult to refine	Some advanced reactors, research uses
Beryllium Oxide	High thermal conductivity, chemically stable at high temperatures	Shares toxicity and cost issues with beryllium	Experimental/advanced compact systems

Reactor Types: How the Pieces Fit Together

The reactor design affects the plant's footprint, cost, operational efficiency, complexity in managing corrosion and fire risks, and ease of waste handling. Choices will depend heavily on where the plant is going and how it will be used. A small rural reactor for power will be designed much differently than a large, multi-purpose commercial reactor.

Pressurized Water Reactors (PWRs) and **Boiling Water Reactors (BWRs)** dominate the global nuclear fleet. PWRs use pressurized water as a coolant and moderator, separating the radioactive primary loop from the secondary steam turbine loop. BWRs allow water to boil directly in the core, simplifying the design but complicating shielding due to radioactive steam.

CANDU reactors use heavy water to achieve high neutron efficiency and allow refueling while the reactor is still running. The dual use of heavy water as moderator and coolant enables the reactor to run on natural uranium, eliminating the need for fuel enrichment. This makes CANDUs attractive for countries with natural uranium or without enrichment infrastructure. However, they are bulky and complex. Their design requires many pressure tubes and heavy shielding components, leading to a larger and more intricate reactor system than compact light-water designs.[5]

Fast Breeder Reactors (FBRs) don't use a moderator to slow neutrons. Instead, they rely on fast neutrons to convert fertile material like U-238 into fissile plutonium.[6] Fast neutrons retain their high energy after fission and are more likely to be absorbed into U-238 without needing to be slowed down. This enables the reactor to breed (produce) more fuel than it consumes, but it also requires higher neutron fluxes and specialized materials that can withstand intense radiation and heat, making FBRs more complex to design and operate than thermal reactors. They offer superior fuel efficiency but require exotic materials and tight controls due to fast neutron reactivity.

Lead-cooled Reactors and Sodium-cooled Fast Reactors (SFRs) are advanced designs that use fast-moving neutrons. This makes them very efficient. They can create more fuel and help burn up long-lasting nuclear waste, which is great for reducing the burden of radioactive materials over time. But these reactors also come with challenges. Sodium reacts violently if it leaks and touches air or water, which can cause fires or even explosions. Lead is much safer in that sense, it's chemically stable, but it's extremely heavy. That weight puts stress on the reactor's structure, and the high temperatures involved mean the materials inside must resist corrosion for long periods.

TerraPower's Natrium Reactor is a current flagship example of SFR technology. It incorporates molten salt but uses scarce HALEU (high-assay low-enriched uranium) fuel instead of thorium for fuel. Backed by both private investment and US Department of Energy funding under the Advanced Reactor Demonstration Program, Natrium combines a sodium-cooled fast reactor core with a molten salt-based thermal energy storage system. This allows the plant to flexibly alternate between steady baseload power or shift to meet peak electricity demands. This innovation could make it an ideal complement to intermittent renewable energy sources. Yet despite its promise, the project has faced real-world challenges. The lack of a commercial domestic supply of HALEU fuel has delayed the Natrium construction timeline, and broader supply chain issues and regulatory uncertainties remain. These setbacks highlight the fragility of advanced nuclear deployment when infrastructure and policy frameworks fail to keep pace with innovation. In this way, TerraPower's Natrium serves as a case in point for the dual nature of SFRs: their enormous potential to revolutionize nuclear energy, and the equally significant technical, regulatory, and economic barriers that must be overcome to realize that potential.[7]

Molten Salt Reactors (MSRs), especially those utilizing thorium, represent a radical departure from traditional nuclear reactor designs. Instead of relying on solid fuel rods housed in pressurized containment vessels, thorium MSRs dissolve fuel directly into a molten fluoride salt mixture that also serves as the reactor's coolant. When thorium-232 absorbs a neutron, it ultimately transforms into uranium-233, a fissile material that can sustain a chain reaction. The configuration allows for continuous fuel reprocessing and removal of waste products, dramatically increasing fuel efficiency and reducing long-lived radioactive waste. The molten salt remains stable at high temperatures and operates at normal atmospheric pressure, avoiding many of the mechanical and safety challenges of high-pressure steam systems used in conventional reactors.[8]

Thorium Molten-Salt Reactor with Heavy Water Moderator

Copenhagen Atomics' Waste Burner mentioned earlier as a nuclear waste consuming reactor, is a single-fluid, fluoride-based molten-salt thermal reactor housed inside a 40-foot shipping container. It can deliver around 100 MW of power while operating at low pressure and high temperature ($\sim$600–700 °C)[9] using a design that features a heavy-water moderator within salt layers, known as the "onion core" design. This approach enables thorium to be converted into uranium-233 fuel while making use of existing nuclear waste as the initial source of fissile material. The company has commissioned two full-scale non-fission prototypes and plans to achieve a full critical test reactor at the Paul Scherrer Institute (PSI) in Switzerland by 2027. They have been collaborating with PSI, a leading nuclear energy research center, to study criticality in thorium molten salt systems. This partnership aims to collect crucial data for licensing molten salt reactors in Europe, with a focus on neutron economy, salt reactivity, and safety measures.[10]

Liquid Fluoride Thorium Molten Salt Reactors (LFTRs)

We've previously discussed liquid fluoride thorium reactors, so let's revisit a LFTR design and compare how they stack up with other thorium reactor designs.

Flibe Energy's LFTR uses a liquid mixture of lithium fluoride and beryllium fluoride salts as both the fuel solvent and coolant in its LFTR. Based on Alvin Weinberg's innovative design, this allows for the continuous conversion of thorium into uranium-233 within the reactor core. LFTRs operate at high temperatures and low pressures, featuring a primary loop that contains the fuel salt and a secondary loop designed to transfer heat efficiently for electricity generation and industrial applications. This setup enhances safety by keeping radioactive materials separate from the power systems and also results in a more compact reactor footprint due to its low-pressure operation. The passive safety features of LFTRs significantly minimize the risk of meltdowns, making these reactors smaller, safer, and well-suited for distributed energy applications and remote locations. Moreover, the unique operating conditions of LFTRs offer additional benefits for processes like desalination, further highlighting their versatility.

Liquid Fluoride Thorium Molten Salt Reactor

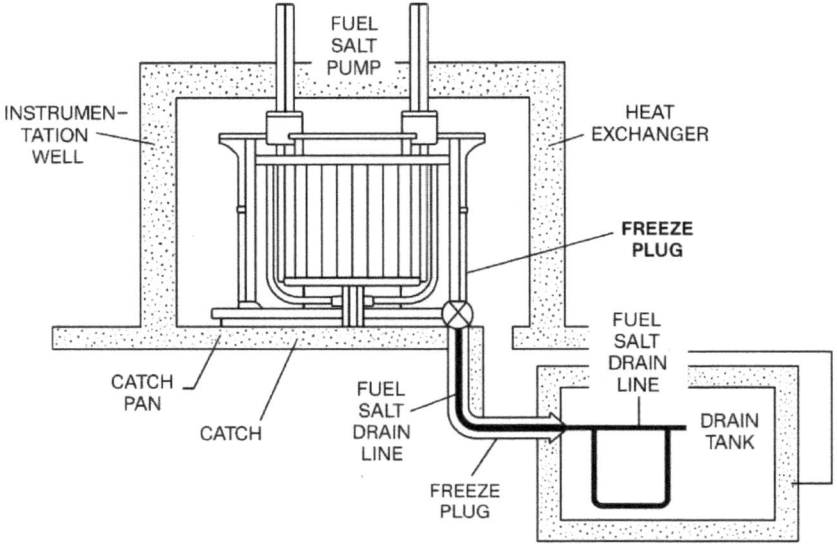

Challenges persist, especially managing corrosion from molten fluoride salts. Advanced materials like Hastelloy-N are needed to maintain the reactor's structural integrity at high temperatures over long periods. While experiments like the Oak Ridge National Laboratory's Molten Salt Reactor Experiment (MSRE) in the 1960s proved the concept, scaling these materials for commercial use presents engineering and cost challenges.[11] Additionally, outdated regulatory frameworks designed for traditional light-water reactors hinder the progress of molten salt technology, resulting in slow innovation and cumbersome approval processes.

Despite the challenges, progress continues around the world. With advances in materials science and policy reforms, molten salt reactors may soon move from experimental status to mainstream energy solutions, offering clean and reliable power while overcoming the limitations of previous nuclear technologies.[12]

Global Momentum: Other Leading Thorium Molten Salt Initiatives

Beyond Flibe Energy and Copenhagen Atomics, a number of ambitious thorium molten salt reactor programs are progressing worldwide, each contributing to the ecosystem of advanced fission innovation.

Moltex Energy, based in the UK and Canada, is developing the Stable Salt Reactor – Wasteburner (SSR-W), which uses molten salt as a coolant and draws on existing spent nuclear fuel as its energy source. While not thorium-fueled in its technology baseline, Moltex's parallel project, the SSR-Th, is explicitly designed to operate with thorium fuel. Both designs emphasize passive safety and cost-effective scalability, with Moltex securing support from New Brunswick Power and the Canadian government for a planned demonstration unit in Point Lepreau, Canada.[13]

China's Thorium-based Molten Salt Reactor (TMSR-LF1) achieved initial criticality in 2023 in Wuwei, Gansu Province. This 2 MW experimental reactor is part of a broader $3 billion strategy to establish grid-integrated thorium MSRs by the early 2030s. Further research is underway to adapt the technology to desert and coastal desalination zones. This most advanced state-sponsored thorium molten salt program is led by the Shanghai Institute of Applied Physics (SINAP) under the Chinese Academy of Sciences.[14]

Year	China's Thorium MSR Program Milestone
2011	Launch of the TMSR program by the Chinese Academy of Sciences (CAS) in Shanghai.
2015	Site preparation begins in Wuwei, Gansu Province.
2018	Construction of the TMSR-LF1 experimental reactor begins.
2021	TMSR-LF1 achieves first criticality using uranium salt fuel.
2023	TMSR-LF1 reaches stable operation; performance data gathered for thorium capability.
2024	Tested thorium-bearing salt fuel blends in static loop experiments.
Apr 2025	First successful demonstration of live fuel addition while reactor remains online.
Mid 2025	Construction begins on 10 MWe second-stage reactor for electricity and hydrogen production.
2029 (planned)	Commissioning of 10 MWe pilot plant.
2030 (planned)	Construction planned for 400 MWt smTMSR-400 commercial reactor.

India's Bhabha Atomic Research Centre (BARC) continues development of the Advanced Heavy Water Reactor (AHWR) using thorium oxide as fuel. While technically a solid-fuel reactor, BARC's long-term roadmap includes liquid-fueled MSR designs that leverage India's vast monazite-derived thorium reserves. India views thorium as a path toward long-term energy independence and has conducted extensive fuel irradiation tests at its Kalpakkam research campus.[15]

The European Molten Salt Reactor (EMSR) initiative, coordinated by EVOL and SAMOFAR consortia, explores both thermal and fast-spectrum thorium MSRs. These academic-industry consortia, backed by the European Commission, focus on fuel salt chemistry, corrosion science, and passive safety modeling, providing critical design input for experimental prototypes.[16]

Together, these efforts suggest a clear trend: thorium molten salt technology is no longer speculative. It is being pursued by companies, consortia, and national labs, each advancing toward the shared goal of clean, compact, and safe nuclear energy.

Comparison of Types of Reactors

Reactor Type	Common Uses	Key Countries Where Used	Advantages	Disadvantages
Pressurized Water Reactor (PWR)	Power generation	USA, France, Japan	Mature technology, widely deployed, strong operational experience	High pressure system, complex containment, long-lived waste
Boiling Water Reactor (BWR)	Power generation	USA, Japan, Sweden	Simplified design, fewer components	Radioactive steam complicates shielding and maintenance
CANDU Reactor	Power generation	Canada, India, South Korea	Uses natural uranium, on-power refueling, high neutron efficiency	Bulky design, complex pressure tube systems, costly heavy water
Fast Breeder Reactor (FBR)	Fuel breeding, waste reduction	Russia, India, France (historically)	Breeds more fuel than it consumes, burns transuranics, closes fuel cycle	Requires exotic materials, high neutron flux, challenging to operate safely
Lead- or Sodium-Cooled Fast Reactor	Fuel breeding, waste burning	Russia (sodium), Europe (lead, R&D)	High neutron economy, reduces long-lived waste	Safety challenges with sodium (fire hazard), corrosion and weight issues with lead
Molten Salt Reactor (MSR)	Advanced power, hybrid uses	China, USA (research), Netherlands	Atmospheric pressure, passive safety, efficient fuel use, ideal for thorium	Corrosion, regulatory lag, immature infrastructure
Liquid Fluoride Thorium Reactor (LFTR)	Power, desalination, remote use	China (testing), USA (concept/research)	Closed fuel cycle, passive safety, continuous fuel reprocessing, clean output	Materials corrosion, licensing barriers, still in prototype/development stage

SMRs: Small Reactors, Big Promise, With Lessons Still Unfolding

Small modular reactors (SMRs) promise to reshape how nuclear power is deployed. With the potential to bring energy to off-grid locations, industrial campuses, and even remote military outposts, SMRs represent a shift toward more flexible, scalable nuclear energy solutions. But as the technology moves from concept to commercialization, it faces promise and pitfalls.

SMR Project Status as of mid-2025

NuScale Power's SMR design was once a leading contender for bringing factory-built, light-water reactors to market. It emphasized safety by using passive cooling systems that rely on natural convection and gravity, rather than mechanical pumps or electrical systems. Full-scale testing showed that the reactors could shut down safely during total power loss. NuScale's vision of modules small enough to be fabricated in a factory and shipped to the site offered reduced construction timelines and the ability to scale by adding reactors incrementally.

However, NuScale's flagship project ultimately collapsed due to a combination of rising costs, waning utility interest, and market pressures. Despite early federal backing and regulatory approvals, the project's estimated costs climbed, casting doubt on its financial competitiveness. Utilities that had signed preliminary agreements pulled back, citing economic uncertainty and insufficient demand.[17]

This outcome does not spell the end for SMRs but rather offers critical lessons. First, cost projections must be more conservative and based on real-world supply chain realities. Second, partnerships with power purchasers must be firm and diversified early in the development cycle. Third, regulatory pathways, while improving, still need streamlining for first-of-a-kind reactors.

Meanwhile, other developers are incorporating these lessons. Rolls-Royce's SMR initiative emphasizes modular construction using proven light-water reactor technology, with factory-built components designed for faster and more predictable deployment. Their approach targets national grids, industrial sites, and export markets seeking carbon-free energy without large conventional reactors' financial and logistical baggage.

The US Department of Defense advanced small modular reactor (SMR) technology through Project Pele, a mobile microreactor design studied to power forward operating bases and other mission-critical infrastructure. The primary goal was to increase energy resilience by reducing the military's reliance on fuel convoys, which have historically been both expensive and vulnerable to attack in conflict zones. Project Pele reactors were designed to generate between 1 and 5 megawatts of power while constructed rugged enough to travel by air, rail, or truck while meeting strict military standards for safety, security, and operational durability.

The design studied for Project Pele used high-assay low-enriched uranium (HALEU) in the form of TRISO fuel—tiny, robust particles with multiple containment layers that can withstand extreme temperatures without melting. Unlike many civilian reactors, Project Pele's design did not require a traditional moderator like water or graphite, since it operates as a fast-spectrum reactor. For coolant, the design incorporates liquid metal, specifically, a heat pipe system using sodium or a sodium-potassium alloy (NaK), both of which are effective at transferring heat efficiently in compact spaces.

However, the choice of coolant also introduced significant challenges. Sodium and NaK are highly reactive with air and water, raising serious fire and safety concerns in the event of a leak; especially problematic for a reactor meant to be mobile and potentially deployed in harsh environments. Additionally, managing heat transport and shielding in such a compact reactor while avoiding coolant exposure to the atmosphere required complex engineering solutions and added to the project's development time and cost.

A thorium-fueled, molten salt moderated and cooled LFTR may yet prove a safer and less complex alternative solution for future field experiments. Despite these hurdles, Project Pele took a pivotal step toward deployable, resilient nuclear energy for defense applications; and raised the possibility of future civilian use of microreactors in remote or emergency settings.[18]

The DoD's initiative echoes the historic path of nuclear submarines, which laid the technological and regulatory groundwork for the commercial nuclear industry in the mid-20th century. By proving the safety and reliability of compact reactors in demanding conditions, Project Pele and similar programs serve as pathfinders for civilian SMR deployment. These defense-backed efforts offer a proving ground insulated from the short-term market pressures undermining NuScale's commercial efforts.

If successful, SMRs could support remote Arctic villages, disaster relief operations, off-grid desalination plants, or island nations with expensive diesel imports. But for now, commercialization remains a cautious climb, requiring realistic expectations, regulatory innovation, and public-private coordination. The vision of small modular reactors is intact, but the path is becoming longer and more complex than initially hoped.

As thorium technology moves from laboratory breakthroughs to real-world deployment, the next question becomes clear: who will step forward to turn this potential into impact, and how can innovators and investors shape the industries of tomorrow?

1. World Nuclear Association, 'Nuclear Fuel Cycle', accessed 2025.
2. INTERNATIONAL ATOMIC ENERGY AGENCY, 'INTRODUCTION TO THE USE OF THORIUM IN NUCLEAR REACTORS', IAEA-TECDOC-1450 (2023).
3. CANADIAN NUCLEAR SAFETY COMMISSION, 'HEAVY WATER AND REACTOR DESIGN CONSIDERATIONS', 2023.
4. US DEPARTMENT OF ENERGY, 'MOLTEN SALT REACTORS', OFFICE OF NUCLEAR ENERGY FACT SHEET, 2024.
5. Canadian Nuclear Safety Commission, 'Heavy Water and Reactor Design Considerations', 2023.
6. NUCLEAR ENERGY AGENCY, 'ADVANCED NUCLEAR FUEL CYCLES AND RADIOACTIVE WASTE MANAGEMENT', 2024.
7. TERRAPOWER. "TERRAPOWER STATEMENT ON THE NATRIUM REACTOR PROJECT." TERRAPOWER NEWSROOM, NOVEMBER 2023
8. GENERATION IV INTERNATIONAL FORUM, 'MOLTEN SALT REACTOR TECHNOLOGY OVERVIEW', UPDATED 2023. COPENHAGEN ATOMICS. "WASTE BURNER REACTOR." COPENHAGEN ATOMICS, ACCESSED JUNE 2025. HTTPS://WWW.COPEN-HAGENATOMICS.COM/WASTEBURNER
9. TERRELL, JEFF. "COPENHAGEN ATOMICS: THE DANISH STARTUP BUILDING MASS-PRODUCED THORIUM REACTORS." MEDIUM, MARCH 5, 2023.
10. TERRELL, JEFF. "COPENHAGEN ATOMICS: THE DANISH STARTUP BUILDING MASS-PRODUCED THORIUM REACTORS." MEDIUM, MARCH 5, 2023.
11. OAK RIDGE NATIONAL LABORATORY, 'MOLTEN SALT REACTOR EXPERIMENT FINAL REPORT', 1972.
12. TERRELL, JEFF. "COPENHAGEN ATOMICS: THE DANISH STARTUP BUILDING MASS-PRODUCED THORIUM REACTORS." MEDIUM, MARCH 5, 2023. HTTPS://MEDI-UM.COM/@TERRELLJEFF/COPENHAGEN-ATOMICS-THORIUM-VISION
13. MOLTEX ENERGY. "STABLE SALT REACTOR – WASTEBURNER." ACCESSED JUNE 2025. HTTPS://WWW.-

MOLTEXENERGY.COM/TECHNOLOGY/SSR-
WASTEBURNER

14. CHINESE ACADEMY OF SCIENCES. "EXPERIMENTAL
    THORIUM MSR REACHES INITIAL OPERATION IN
    GANSU." CAS BULLETIN, OCTOBER 2023.

15. BHABHA ATOMIC RESEARCH CENTRE. "THORIUM FUEL
    CYCLE IN INDIA: TECHNOLOGY STATUS AND FUTURE
    ROADMAP." BARC ANNUAL REPORT, 2023.

16. EVOL/SAMOFAR. "EUROPEAN RESEARCH ON MOLTEN
    SALT REACTORS." EUROPEAN COMMISSION HORIZON
    2020 REPORTS, 2022–2024.

17. World Nuclear News, 'NUSCALE POWER PROJECT
    CANCELLED OVER COST CONCERNS', 2024.

18. US DEPARTMENT OF DEFENSE, 'PROJECT PELE: MOBILE
    MICROREACTOR DEVELOPMENT UPDATE', 2025.

Chapter 5

Powerful Possibilities: A Call for Innovators and Investors Across Industry Needs

A uthor's remarks:

My perspective on practical innovation is shaped by a series of experiences that, in retrospect, come together neatly. As mentioned previously, I first encountered the quiet reliability of nuclear power as a young mechanic in the Navy's submarine force, where the steady heat from a reactor ensured life support, propulsion, and freedom from the fuel convoys that burden surface ships. After receiving my ROTC commission, I served as the repair-division officer on a destroyer, where the habitability systems influenced daily life, and a desalination plant transformed cloudy seawater into clear, drinkable water.

At the Navy Damage Control School, I taught about the fire triangle, which includes oxygen, heat, and fuel: remove any of these, and the fire goes out.[1] We also covered the fire tetrahedron, which adds a fourth element: the chemical chain reaction, where catalysts help sustain combustion. Inhibiting this chain reaction also puts out a fire.

Upon leaving active duty and transitioning to aerospace systems engineering with a master's degree in technology management, I developed the concept of an innovation fire triangle. I realized that every successful technology requires the oxygen of novelty, a steady stream of funding to fuel it along towards development, and a market that burns hot enough with customer demand to keep it alive. The innovator acts as the catalyst, bringing all these elements together in an explosive mix of ideas and action.

Every breakthrough I've seen, from high-temperature materials to autonomous vehicles, has met these critical requirements. Remove any side of the triangle, and the innovation inevitably flames out.

The Innovation Fire Triangle

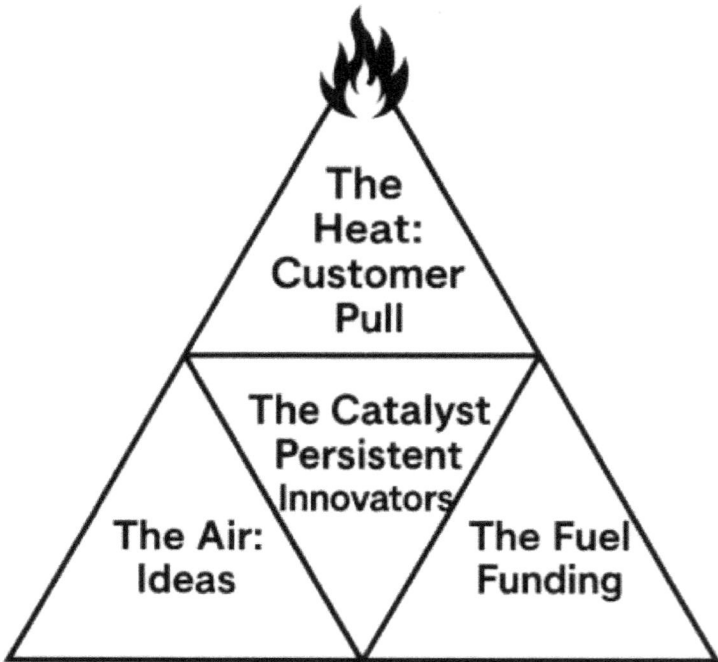

The Innovation Fire Triangle

The
Heat:
Customer
Pull

The Catalyst
Persistent
Innovators

The Air:
Ideas

The Fuel
Funding

This chapter will apply the innovation fire-triangle framework and highlight *what* tech is novel, *where* markets are heating up, *how* to fuel it with funding, and *who* the innovators are who serve as the catalysts to make it happen.

Thorium represents a technological enabler capable of sparking innovation across multiple industries—if someone is persistent enough to light it off. Let's explore the potential boom in economic growth across sectors as diverse as AI, energy, fuel, medicine, agriculture, defense, transportation, and beyond.

The Oxygen of Novelty: Emerging Technologies

The Rise of Small Modular Reactors – and Microreactors

Small modular reactors (SMRs) are factory-built power modules that generate up to 300 MW, which is about one-third of what a typical large power plant generates.[2] Whether cooled by water, gas, liquid metal, or molten salt, each module is shipped to site, assembled quickly, and scaled as needed, slashing construction time, capital risk, and regulatory burden for rural grids, mines, desalination plants, or industrial parks.

Microreactors push the idea further. Packing 1-10 MW into a truck- or rail-portable unit, they can power forward bases, disaster zones, Arctic villages, or lunar outposts with minimal staffing.[3] Early models use high-assay low-enriched uranium (HALEU) fuel; future versions can run on thorium-based liquid fuel.

Thorium couples naturally with this architecture. Molten-salt reactors (MSRs) that dissolve thorium fuel operate at normal atmospheric pressure and high temperature, eliminating bulky containment domes and long cooling loops. Lower mechanical complexity means lower first-of-a-kind cost, an advantage for small island nations or remote regions.[4]

NuScale's pressurized-water SMRs faced technical and economic challenges due to high-pressure operation, water dependency, and complex safety systems. Despite early enthusiasm, several projects, including NuScale's flagship effort, were eventually canceled due to rising costs, regulatory friction, and minimal market uptake.[5][6][7]

In comparison, thorium MSRs avoid high-pressure hazards and water dependence. Their FLiBe salt coolant stays chemically stable across a wide temperature range, avoiding the sodium–air reactivity of fast-reactor concepts.[8]

Flibe Energy[9] and Kairos Power[10] in the United States and programs in Canada, the UK, and China are now moving thorium-compatible SMRs from prototype to licensing.[11] China has already operated an experimental thorium MSR and targets commercial rollout in the 2030s.[12]

If these efforts succeed, thorium-fueled SMRs and microreactors could deliver a new class of clean, compact, and widely available energy to fill the gaps left by older water-based reactor systems and supply power everywhere from tech hubs near cities to remote, hard-to-reach areas.[13]

A Nuclear Power that Eliminates Nuclear Waste

Imagine a world where nuclear waste is a resource rather than a threat. Thorium liquid fluoride thorium reactors (LFTRs) offer a practical path toward that future. The uranium nuclear fuel cycle is inefficient, producing long-lived radioactive waste that requires secure storage. LFTR technology can consume spent nuclear fuel and transform it into valuable electrical power, addressing the challenge of eliminating the approximately 2,000 metric tons of spent fuel generated annually in the US.[14]

LFTRs would eliminate spent uranium waste and generate electricity, providing operators with multiple revenue streams. As explained earlier, excess energy can be utilized for the co-production of other valuable resources, such as freshwater, industrial salts, synthetic fuels, and medical isotopes. This multi-purpose approach presents a compelling business model while helping solve important environmental and energy challenges.

How can innovators and advocates help turn nuclear waste into a new fuel source? By developing new chemical processes to use spent uranium fuel and safely handle radioactive waste efficiently. We also need to improve systems that automatically handle fuel and separate waste, allowing us to recycle nuclear waste into energy while reducing safety risks and meeting regulations. Additionally, we need to create materials for the thorium reactors that can resist damage from molten salts and high radiation. By investing in LFTR development, we can transform nuclear waste into a valuable asset for sustainable energy. Here's an example of a company leading the way:

Copenhagen Atomics Waste Burners are molten salt reactors compact enough to fit inside a standard 40-foot shipping container, but powerful enough to produce around 100 megawatts of thermal energy at high temperatures and low pressure.

What sets them apart is their ability to consume nuclear waste as part of the fuel mix while using thorium to generate clean energy. Their "onion core" reactor design allows layers of different materials to work together efficiently. The waste gets burned down, and the thorium is gradually converted into usable nuclear fuel. That makes the system both productive and self-cleaning.

Copenhagen Atomics is a Danish startup founded in 2014 with a bold vision for the future of clean energy, and their business model isn't just to sell reactors. They plan to finance, own, and operate fleets of them, eventually producing thousands of units per year. Their goal is to deliver electricity for under $20 per megawatt-hour, which is far below the current global average. [15]

But electricity is just the beginning. In Indonesia, they're developing projects to produce clean ammonia using reactor heat—a process that

89

supports fertilizer production without carbon emissions.[16] By combining waste reduction, electricity generation, and chemical production, they're building a model that turns multiple challenges into multiple revenue streams. It's a powerful example of how next-generation reactors can solve several problems at once and make a profit doing it.

Chemical Processing of Fertilizers, Fuels, and High-Tech Materials

Thorium's value in industry goes beyond making electricity. In powder or oxide form, it can act as a catalyst, a substance that speeds up chemical reactions without being depleted. Because thorium stays solid and stable at very high temperatures, it lets factories run hotter and longer, saving energy and cutting waste.

Fertilizers. The global food supply depends on ammonia, the base ingredient for most nitrogen fertilizers. Today's Haber-Bosch plants must heat and pressurize nitrogen and hydrogen until they combine, using about two percent of the world's total energy. Thorium oxide holds up better than many standard metal catalysts under these harsh conditions, so that the same plant can make the same ton of ammonia with less heat, less pressure, and less fuel. This application and the Haber-Bosch link to thorium is speculative until innovators prove it out.

Fuel refining. Let's talk about fuel refining: the process of turning thick, heavy crude oil into lighter fuels like gasoline, diesel, or jet fuel. This transformation depends on something called "cracking," which means breaking apart long hydrocarbon chains using high heat. This process is particularly tough on equipment. It requires extremely high temperatures, and over time, the catalysts (the materials that make the reaction work) start to wear out. That's where thorium has a key advantage. Thorium-based catalysts can handle those high temperatures for much longer before they need to be replaced. That means the refinery runs more efficiently, processes more fuel in less time and burns less gas to keep the heat up. The result? Lower operating costs and fewer carbon emissions. So even in industries built on fossil fuels, thorium has the potential to make things cleaner, cheaper, and more sustainable.

Advanced materials. Thorium's high melting point and chemical stability also help when making specialty glass for camera lenses, heat-resistant ceramics for jet engines, or superconducting wires for medical scanners. A small dose of thorium compound can keep the melt uniform and free of bubbles, leading to stronger parts and less scrap.

None of these benefits happen automatically; engineers still need to design equipment that handles thorium safely and prevents dust release. Universities, national labs, and companies are beginning that work now, testing thorium catalysts in pilot reactors and sharing data on performance and environmental safeguards. If those trials scale up, thorium could give fertilizer makers, refiners, and high-tech manufacturers a cleaner, cheaper way to run the reactions that support modern life.

Novelty alone isn't enough; without customers the flame dies, so let's turn to where there is a strong market demand for this new technology.

The Heat of the Market Pull: Where's the Demand

AI's Insatiable Energy Appetite and the Nuclear Solution

Skyrocketing electrical demand for artificial intelligence (AI) data centers presents a near-term entry point for bringing thorium reactors online. The rapid advancement of AI technologies has led to an unprecedented surge in energy consumption. Data centers, the back-bone of AI operations, are projected to consume more electricity than entire nations. According to the International Energy Agency (IEA), global electricity demand from data centers is expected to double by 2030, with AI-specific centers expected to quadruple consumption.[17]

In the United States, data centers are on course to account for almost half of the growth in electricity demand between now and 2030. This surge has prompted several US states to explore advanced nuclear reactors as a viable solution to meet the growing energy needs.

At the federal level, the urgency has been formally recognized: in May 2025, US Presidential Executive Order 14090 directed the Department of Energy to designate AI data centers as critical defense infrastructure and to fast-track the deployment of advanced reactors, including molten salt designs, to power these facilities within 30 months. This policy link between AI resilience and advanced nuclear deployment adds momentum and priority to thorium-powered reactor strategies.

Three U.S. states show different challenges with their power grids and how they are responding with nuclear energy.:

Pennsylvania: Amazon bought a data center that connects directly to TalenEnergy's Susquehanna nuclear power plant.[18] This shows a growing trend of placing data centers near nuclear facilities. While the Federal Energy Regulatory Commission (FERC) turned down a request to increase power use beyond 300 megawatts[19], this move highlights how the tech industry is interested in using nuclear energy to support AI infrastructure.

Michigan: The state is investing nearly $2 billion to reopen the Palisades Nuclear Plant, which closed in 2022. Slated to resume operations in October 2025, the plant significantly boosts the 8 energy capacity, which could support growing AI data center demands.[20]

Virginia: Northern Virginia, a hub for data centers, faces a seven-year wait for new facilities to connect to the electrical grid due to overwhelming demand. This bottleneck has led to discussions about integrating SMRs to provide localized, reliable power sources for data centers.[21]

How Reactors May Benefit from the AI Data Centers They Power

Integrating AI into nuclear energy operations enhances the efficiency and safety of reactors, especially molten salt reactors. Researchers at Argonne National Laboratory have developed AI tools to optimize reactor design and enable real-time monitoring and diagnostics. This reduces the need for human oversight and improves reactor responsiveness.[22]

AI also aids in managing the complex chemical process of MSRs, speeding up fuel optimization and improving safety. Because there are so many possible molten salt mixtures, testing them all would be expensive and time-consuming. AI models can help by predicting the properties of different salt mixtures, making it easier to determine the best candidates for reactors.[23]

Advanced small modular reactors, especially molten salt designs, are now being paired with something called a digital twin. Think of it as a real-time virtual copy of the reactor that runs in parallel with the physical one. This digital twin uses live sensor data from the reactor to simulate what's happening inside, moment by moment. It can spot issues before they turn into problems, helping operators plan maintenance before anything breaks. It also allows for smarter, more autonomous control and can adjust power output dynamically based on demand. In short, it's like giving the reactor a second monitor that watches everything, learns as it goes, and helps keep things running safely and efficiently.[24]

For molten salt reactors in particular, where the fuel is a flowing chemical mixture, AI tools can continuously monitor salt composition, detect anomalies, and optimize performance.

These smart systems reduce the need for constant human oversight, making microreactors safer and more scalable, especially in remote or high-security applications like data centers, desalination plants, or military bases. As AI grows more powerful, it may become not only the reason we need more nuclear power, but the means by which we safely manage it.

The symbiotic relationship between AI and advanced nuclear reactors addresses the challenges of meeting AI's energy demands while improving nuclear operations. AI will play a role in improving energy generation and management, simultaneously sustaining future AI digital infrastructure.

Mission-Critical Power: Defense, Disaster, and Harsh Environments

Energy logistics often present considerable challenges for modern military operations. Supplying fuel to remote bases can be more perilous than actual combat. However, thorium reactors could enable these bases to generate their electricity independently and even produce water from local resources. This could eliminate the need for fuel convoys, which are costly and vulnerable to attack. A microreactor could restore power and water supplies within hours instead of days.

Energy independence is a mission imperative in today's military landscape. Thorium systems would benefit not only remote bases but also mobile hospitals, emergency response teams, and communities affected by disasters. In the critical hours and days following a natural disaster or humanitarian emergency, conventional power grids are often down, fuel supply chains are disrupted, and vital services are compromised. A thorium-powered microreactor— compact, self-contained, and transportable by truck, rail, or cargo plane—could provide a stable, decentralized energy source for essential public services.

Hospitals and emergency care units require constant, high-quality power for ventilators, surgical lighting, medical imaging devices, and intensive care equipment. Equally vital is the continuous operation of medical refrigeration systems, which preserve vaccines, blood plasma, insulin, and biologic medicines. A thorium microreactor operating at 1 to 5 megawatts could supply uninterrupted power to these high-priority needs for months, without the need for diesel fuel convoys or battery replacement.

Beyond medical care, thorium reactors can sustain lighting and temperature control for emergency shelters, water purification systems using reverse osmosis, and mobile sanitation units, all of which are critical for preventing secondary public health crises. Transportation hubs, such as makeshift airfields, landing zones, and electric vehicle charging stations, could also be powered reliably, helping to coordinate inbound humanitarian aid. Satellite uplinks and radio repeaters, vital for communication between field teams and command centers, depend on continuous, high-stability energy sources that batteries alone often cannot provide.

Thorium SMR's advantage lies in its long-duration, clean operation, making it ideal for prolonged recovery efforts where weather-dependent renewables or fuel deliveries are unreliable. By removing dependence on fragile fuel logistics and extending the operational reach of disaster response units, thorium microreactors offer a transformative tool for climate resilience, emergency response, and humanitarian relief missions alike.

In addition to powering essential functions, these reactors could enhance operational efficiency by powering reverse osmosis systems to purify nearby water sources. This capability enables military forces to operate with greater autonomy and resilience, reducing logistical burdens associated with fuel and water supply chains. Mobile thorium-powered units would further improve operations at forward operating bases, support ground missions, and aid disaster relief efforts. Designed for quick deployment in field conditions, these compact reactors ensure a stable power source even in harsh environments and are particularly suitable for remote installations lacking traditional power infrastructure.

Thorium reactors also hold promise for advanced military technologies, including directed-energy weapons, which require large amounts of energy. This opens new possibilities for innovative defense systems that are unconstrained by power availability.

A further advantage is what happens when the mission ends. Because a thorium molten-salt core does not breed weapons-grade plutonium and its uranium-233 stream is laced with hard-gamma uranium-232, the fuel is unattractive for diversion. The US Department of Defense's Project Pele notes that a mobile MSR "would be suitable for handover to civil authorities once tactical requirements are met," provided routine safeguards remain in place. In practical terms, a forward base could leave its reactor behind as a community micro-grid, delivering steady power and clean water long after the troops depart. Such transfers would replace the usual diesel generators with their constant fuel bills and emissions with a self-contained plant designed to run for decades, turning a short-term deployment asset into a foundation for local recovery and economic growth.

Thorium technology enables energy flexibility and independence within the defense sector, reduces reliance on fossil fuels and fuel transport, and supports environmental sustainability by lowering carbon emissions. This versatility extends to maritime operations, where smaller thorium reactors could serve as next-generation propulsion systems, replacing fossil fuels and eliminating the need to refuel naval vessels smaller than nuclear aircraft carriers.

By using thorium reactors, the defense industry can transform how military operations and disaster relief are powered, leading in responsible innovation that benefits both security and local communities. In later chapters we'll look at applying these mobile solutions to other harsh environments in remote areas globally and in space.

Energy-Intensive Cryogenic Cooling

One high-value application often overlooked in discussions of energy access is cryogenic cooling. Cryogenic cooling is used to store liquefied gases like oxygen, nitrogen, and helium. It underpins everything from vaccine preservation to semiconductor manufacturing and even quantum computing. These systems are exceptionally energy-intensive, particularly when operating at scale or in remote environments where power is unreliable or costly.

A thorium-powered small modular reactor, offering stable, high-temperature heat and reliable electricity, could dramatically improve the availability and affordability of cryogenic systems by co-locating power generation with cooling demand. For instance, liquid oxygen (LOX) and liquid nitrogen (LIN) production in rural hospitals, semiconductor fabs, or even spaceports could be paired with SMRs to avoid costly fuel imports and grid failures.[25]

Moreover, thorium's ability to simultaneously generate electricity and process heat enables efficient integration with Stirling-cycle or helium-expander cryocoolers.[26] This means a thorium reactor can power both the machines that keep things very cold and the systems that use that cold to store important materials like medicine, gas, or computer parts, all at the same time, without needing extra equipment or fuel. This kind of energy-density advantage is especially relevant as emerging technologies from gene therapies to superconductors scale into regions where cryogenics has traditionally been out of reach.[27]

Power-hungry Agriculture

In arid regions and megacities, vertical farms require round-the-clock power for LED lighting, nutrient cycling, and automated irrigation. Solar may help, but only nuclear can deliver consistent, scalable energy without massive land use. A thorium reactor could make vertical farming and enclosed agriculture viable where food and energy insecurity intersect.

In Chapter 8 we will take a deep look at farming, which is full of energy demands. In agriculture, sustainability is more important than ever. As populations grow and climate change affects farming, we need ways to produce food responsibly. Thorium power can drive irrigation pumps and controlled-environment farms, cutting agriculture's carbon footprint.

Ocean Transport: Clean Energy for the Shipping Sector

Shipping accounts for over 80% of global trade and emits more than 3% of global carbon dioxide. While synthetic fuels may be a partial solution, thorium-fueled molten salt reactors offer a long-term alternative for marine vessels. Unlike diesel, thorium produces no carbon or sulfur emissions, and it does not require a costly new infrastructure for refueling, like hydrogen or ammonia.

These thorium reactors could enhance a ship's range, reduce downtime, and eliminate refueling in hostile areas, leading to higher profitability and lower insurance premiums due to reduced pollution risks. As stricter emissions regulations increase costs for fossil fuels, thorium's advantages, such as improved energy efficiency and decreased reliance on volatile fuel prices, will make it an attractive investment.

While maritime nuclear propulsion has been used by military vessels for decades, commercial applications have been slow to adopt. However, advancements in compact reactor design and regulatory reform could change this. The concept of a sealed thorium SMR for container ships or tankers holds potential, but development has yet to occur, despite earlier proposals from Alvin Weinberg in the 1960s.

Integrating thorium molten salt reactors into commercial ships presents challenges, including miniaturization, efficient heat management, and advanced safety systems for harsh maritime conditions. Regulatory approval and specialized crew training are also critical for implementing nuclear propulsion in the shipping industry. Collaboration among innovators, engineers, and regulators will be essential.

Successful uranium-fueled precedents, such as Russian nuclear icebreakers and US Navy ships, provide a blueprint for adapting thorium reactors to commercial shipping. And Core Power, a UK startup, is already taking on the challenge of developing a marine molten salt reactor for bulk shipping. By leveraging these examples, we can move towards a sustainable future in maritime transport and meaningfully reduce global fossil fuel dependence.

Industries with Immediate Demand for MSR Heat

Industry Sector	Primary Heat Use	MSR Advantage	Example Applications
Desalination	Evaporation, membrane pretreatment	Stable thermal and electric output, efficient waste heat reuse	RO + MED hybrid plants, Zero-Liquid Discharge
Fertilizer Production	Ammonia synthesis via hydrogen production (Haber-Bosch)	High-temperature hydrogen generation from thermochemical water splitting	On-site green ammonia for agriculture
Oil & Gas / Petrochem	Steam reforming, distillation, cracking	Decarbonized heat source, retrofittable into brownfield facilities	Hydrogen-from-hydrocarbon upgrade, flare gas use
Synthetic Fuels	Fischer-Tropsch synthesis, CO_2 reduction	High process heat and CO_2 availability via direct air capture + electrolysis	Carbon-neutral diesel, kerosene, marine fuel
Mining & Materials	Ore reduction, process heating, rare earth separation	Off-grid thermal and electrical integration; radiation-hardened process designs	Lithium extraction, aluminum production
District Heating	Seasonal heating for residential/industrial buildings	Steady low-carbon thermal supply with potential co-generation	Northern Europe, East Asia urban heat grids
Pulp & Paper	Steam for bleaching, drying, and pulp digestion	Modular heat integration with biomass or recycled inputs	Closed-loop water reuse and effluent treatment
Data Centers	Cooling, dehumidification, power-intensive compute demand	Co-located SMR can power and manage thermal loads with low-emission footprint	AI training farms, edge compute with water reuse
Medical Isotopes	Target irradiation and production of molybdenum-99, lutetium	Neutron flux access and modular co-location with hospitals or production hubs	Regional isotope manufacturing

Advanced Applications: Energy Where Innovation Happens

While most discussions around nuclear energy rightly focus on critical infrastructure like water, lighting, and emergency power, a new category of high-value applications is emerging. These applications are so power-hungry and strategically vital that they demand a reassessment of how and where energy is delivered. Thorium-powered reactors, especially small modular and molten salt designs, offer a unique advantage in these frontier use cases.

Semiconductor Manufacturing and Precision Fabs: Cutting-edge fabs depend on ultra-reliable power. Lithography, plasma etching, and chip inspection processes all demand stable voltage and high throughput. A thorium SMR co-located with semiconductor facilities could ensure uninterrupted baseload power, reduce dependency on diesel backup, and protect national chip-making efforts from grid outages.

Advanced Ceramics and Metallurgy: Industries that produce turbine blades, rare earth magnets, and aerospace ceramics rely on process temperatures well above what standard grids can support efficiently. Thorium reactors operating at 600–750°C could serve as on-site power and heat sources for facilities that produce the building blocks of modern infrastructure and defense.

Desalination Coupled with Mineral Recovery for Electric Vehicle Batteries: The brine produced from desalination is often treated as waste, but it actually contains dissolved minerals that hold strategic value, such as magnesium, lithium, cobalt, and even small amounts of rare earth elements. Desalination facilities powered by thorium could also run advanced electrochemical systems that extract important battery materials from seawater and salty waste. This turns what was once seen as a waste disposal issue into a valuable economic opportunity.

In addition to seawater mining, SMRs powered by thorium can provide the high-temperature, stable energy needed for extracting lithium from brines, refining copper in energy-scarce mining regions, and purifying graphite. All of these are critical for building strong supply chains for electric vehicles and batteries. Today, these processes are often limited by the lack of reliable electricity or natural gas. By placing thorium reactors next to desalination plants or mining operations in resource-rich areas that lack infrastructure, such as the Andes, Central Africa, or Australia, we could provide both clean water and dependable energy. This would also support the production of raw materials essential for a sustainable electric transportation future.

Hydrogen Liquefaction and Transport: Producing hydrogen is only half the battle. Storing and moving it can consume up to 30% of its energy content. Liquefying hydrogen to -253°C is required for dense storage and long-distance transport, but it requires a steady, high-capacity energy supply. A thorium SMR could co-locate with green hydrogen facilities to deliver both electricity and process heat, enabling on-site liquefaction and boosting the economics of the hydrogen economy.

Carbon Capture and Mineralization: While low-temperature carbon capture has gained attention, long-term storage through mineralization (transforming CO_2 into stable carbonates) requires sustained thermal input. Thorium reactors could drive this process where it's most needed: near industrial zones or even at sea, turning waste gas into construction-grade stone while closing the carbon loop.

Subsea and Deep-Ocean Systems: From autonomous underwater vehicles to long-term oceanographic monitoring, subsea systems demand energy without sunlight, fuel deliveries, or surface cables. Thorium microreactors, sealed and submersible, could revolutionize undersea exploration and defense.

Emerging markets currently face significant energy challenges, presenting an opportunity for innovative solutions like thorium reactors. These reactors have the potential to create new energy infrastructure that meets essential needs while fostering innovation and growth. Investors and innovators should recognize that thorium's benefits extend beyond basic utilities like clean water and electricity and could play a pivotal role in catalyzing the next industrial revolution.

Too often, great ideas with strong market demand run out of fuel and fizzle out. Technology development can require many funding sources to transition into a fully fielded technology, so next, let's investigate where to find the money.

How to Fuel it with Funding

Medical Isotopes: A Shortage Solved by a New Source

Even the waste byproducts in a thorium reactor can serve as a valuable resource. Certain radioactive isotopes used in cancer treatment and imaging are in critically short supply. Radioactive isotopes play an important role in treating diseases, especially cancer. These isotopes are not common because they have short half-lives, but they are crucial for targeting cancer cells effectively. As liquid fluoride thorium reactors (LFTRs) operate, they produce byproducts like iodine-131, which helps treat thyroid cancer, and thorium-229, which changes into bismuth-213. Bismuth-213 has applicability for cancer treatment. When linked to antibodies, it can target cancer cells directly, delivering harmful alpha radiation to them while protecting healthy tissue nearby. This focused method increases the chances of killing cancer cells and reducing damage to normal cells. The quick decay of these isotopes also makes them safer, as they lose their radioactivity quickly and lower long-term exposure risks.

The market potential for byproducts from thorium reactors in the medical sector is sizeable but largely untapped. Currently, the production of medical isotopes is limited, making them expensive and hard to obtain. LFTRs can serve as a reliable source for these rare isotopes, which are in demand for targeted cancer therapies. These reactors could streamline production and improve availability by directly generating these isotopes.

New technologies are needed to extract and purify isotopes from reactor fuel salt so that byproducts from thorium reactors can be used in medical applications. Engineers and scientists must collaborate to create efficient systems that adhere to strict medical standards, while also establishing regulatory frameworks for safety. Partnerships with medical institutions and regulatory bodies are necessary to overcome these hurdles.

Currently, producing medical isotopes is complex and costly because it relies on a few large research reactors, often located overseas. This leads to delays, supply chain bottlenecks, and geopolitical risks. That means hospitals often face delays, limited availability, and supply chains that stretch across continents. On top of that, the cost of transporting these sensitive materials is high, especially because some isotopes decay quickly and need cold storage the entire way.

Thorium-fueled microreactors could change this scenario. Small, portable reactors producing around 10 megawatts of power could be set up right where the isotopes are needed. Imagine one on the campus of a major hospital or at a regional medical hub. Instead of flying in isotopes from abroad, they could be produced on-site, reliably and consistently. That means no more scrambling for scarce shipments, no more sky-high transport costs, and no more racing against the clock before the isotope decays. These microreactors would simplify logistics and could dramatically lower the cost of producing life-saving treatments, especially for cancer patients who depend on precision-targeted radiation therapy.

Moreover, molten salt microreactors with thorium fuel offer benefits like online reprocessing and isotope extraction without reactor shutdowns. Their modular designs enable standardization and scalability to meet growing demands in nuclear medicine. This approach shifts isotope production from a centralized model to a distributed network, making advanced cancer treatments more accessible, particularly in areas with limited nuclear infrastructure.

TerraPower provides one of the most promising commercial efforts to address this shortage. The company is actively scaling production of Actinium-225 (Ac-225), a rare alpha-emitting isotope with powerful applications in targeted cancer therapies.[28] Actinium-225 is particularly useful in treating late-stage cancers by attaching to antibodies that target malignant cells, then delivering a lethal dose of radiation with minimal damage to surrounding tissue. TerraPower is investing in domestic production through its subsidiary, TerraPower Isotopes, collaborating with both national labs and the Department of Energy to ensure a steady and scalable supply.[29]

In doing so, TerraPower is building a diversified revenue model by combining isotope production with advanced reactor development, such as the Natrium fast reactor. It is also exploring how to use these reactors to supply heat and electricity for data centers, synthetic fuels, and industrial processes. This stacked-value approach, which combines energy, isotopes, and industrial applications, reflects a similar strategy being explored for thorium reactors and highlights how nuclear innovation can be economically viable beyond just selling electricity.[30]

Flibe Energy has also made significant strides in medical isotope research using thorium-based molten salt reactor technology. In partnership with Oak Ridge National Laboratory and university researchers, Flibe has demonstrated how their reactor fuel cycle can produce high-value isotopes from thorium-229, which decays into bismuth-213, a potent alpha emitter used in targeted cancer therapies. Their experimental efforts include developing salt-processing methods that selectively extract these isotopes during reactor operation, without requiring a full shutdown. By building extraction capability directly into the LFTR design, Flibe is positioning thorium reactors as not only clean energy sources but also steady producers of life-saving medical materials. This work supports a future where thorium-powered microreactors could supply isotopes on demand at hospitals, regional medical centers, or even battlefield clinics, bypassing delays and supply bottlenecks.

Synergies with Renewables and Hydrogen

Thorium small modular reactors are a good match for renewable energy sources like wind and solar. While renewable energy is clean, it can be unreliable; sunlight and wind aren't always available when we need them. Thorium reactors provide a steady flow of energy, helping to keep the power grid stable and reducing the need for fossil fuels. This partnership extends to green hydrogen, which many countries are turning to for tough-to-decarbonize industries like steelmaking, aviation, and heavy transport.

One effective way to produce hydrogen on a large scale is through the sulfur iodine cycle. This complex process uses heat to split water into hydrogen and oxygen with chemicals that include sulfur and iodine. However, it needs very high temperatures above 700 degrees Celsius. Most traditional energy sources can't reach these temperatures, but thorium molten salt reactors can, running at between 600 and 800 degrees Celsius. This makes them suitable for advanced hydrogen production that regular renewables cannot handle alone.

The sulfur iodine cycle only requires heat to split water, not electricity. This means a thorium reactor can produce hydrogen without needing a lot of power. It is a thermal process that works well for large-scale hydrogen production at industrial sites. For instance, hydrogen made this way could be sent to green steel plants, mixed into existing gas lines, or turned into liquid fuels like ammonia or methanol for shipping and air travel. The U.S. Department of Energy sees this process as a perfect fit for next-generation reactors, and thorium designs are among the top options.

There is another intriguing possibility: solid oxide electrolysis (SOE). This technique splits water into hydrogen and oxygen using both electricity and heat. When paired with a thorium reactor, it becomes significantly more efficient than standard electrolysis. It's like getting more hydrogen from the same energy input because we're using both the electricity and the thermal energy the reactor already produces. Various projects in Japan, China, and the European Union are exploring advanced nuclear-hydrogen integration models, with thorium MSRs being evaluated as a key technology. In short, thorium reactors don't compete with renewables; they complement them. They fill in the gaps, power the industrial backbone, and help build a cleaner, more resilient energy system.

Unlocking New Markets Through Abundant, Clean Power

Energy transitions have consistently transformed economies. Small modular molten salt reactors fueled by thorium have the potential to provide stable electricity and industrial heat to areas previously without reliable power, such as remote mining sites, rural farms, desalination plants, and AI data centers.

The growing issue of global water scarcity adds another layer to this economic transformation. The UN estimates that by 2050, over 5 billion people will face water shortages. A 2022 study by the International Atomic Energy Agency (IAEA) reveals that integrated nuclear desalination systems, particularly those utilizing advanced molten salt reactor designs, can significantly lower the cost of freshwater.[31] These systems become even more effective when paired with a dependable heat and power source like thorium.

The World Bank has noted that every dollar invested in climate-resilient water infrastructure can generate four dollars in economic returns through enhanced health, productivity, and agricultural output.[32] Research by Hargraves indicates that molten salt reactors using thorium could not only reduce energy and desalination costs but also deliver a favorable return on investment within two decades, provided the technology is scaled effectively.[33]

Thorium also presents a cost-effective means to modernize aging power plants without the need for extensive overhauls. This adaptability allows for a smoother transition to this innovative energy source, as thorium reactors have lower operational costs and increased energy efficiency. The combined economic and environmental advantages make thorium an appealing option for countries aiming to decrease their carbon footprint while ensuring energy security.

By offering abundant, localized power, thorium reactors can lower energy costs for industries, stimulate regional development, and support sectors like synthetic fuels, green hydrogen, and climate-resilient agriculture. According to a 2021 article from the International Energy Agency, decentralized clean energy systems are projected to spur over $10 trillion in global infrastructure investment by 2050, with modular reactors playing a critical role.

TO RECAP, THORIUM REACTORS CAN PRODUCE A VARIETY OF
VALUABLE RESOURCES FOR DIFFERENT SECTORS, INCLUDING:

- ELECTRICITY FOR HOMES, BUSINESSES, AND DATA
  CENTERS
- MEDICAL ISOTOPES FOR HEALTHCARE
- FRESHWATER THROUGH DESALINATION
- SYNTHETIC FUELS FOR TRANSPORTATION AND
  AGRICULTURE
- AMMONIA FOR CLEAN FERTILIZERS

This multi-product strategy enhances the financial stability of thorium-powered facilities. The US Department of Energy's 2022 Advanced Reactor Demonstration Program affirmed the importance of generating diverse products to ensure the profitability of future nuclear plants. While the initial investment in thorium reactors may be considerable, the potential for multiple revenue sources promises sizeable returns.

Stacking the Returns: A Layer-by-Layer Business Case for Thorium SMRs

Selling electricity and medical isotopes: Let's start with a straightforward example: a 50-megawatt thorium molten salt reactor, feeding electricity into a local power grid and possibly a nearby AI data center. That's a small reactor by utility standards, but big enough to power a town or help run energy-hungry servers.

According to estimates from the U.S. Advanced Reactor Demonstration Program, building one of these would cost about $200 million. That's based on a first-of-its-kind price of around $4,000 per kilowatt of capacity.[34]

Now let's look at revenue. If the reactor runs most of the year, about 8,000 hours, and sells power at 70 dollars per megawatt-hour, it would bring in roughly $28 million a year from electricity alone. The cost to operate and maintain the reactor comes to around $4 million annually, which leaves about $24 million in net income before debt or financing. That means the reactor could pay for itself in about eight to nine years, just from selling electricity.

But here's where it gets interesting. If the reactor includes a special side loop—something designed to extract medical isotopes—it can do even more. For example, it could produce bismuth-213, a powerful isotope used in targeted cancer therapy. In North America, the market price averages around $8,000 per millicurie. A 50-megawatt reactor could make roughly 900 curies a year.

Even after covering the cost of the hot cell, essentially the high-security room used to handle radioactive materials, that could add another $6 to $7 million in yearly revenue. With that bonus income, the reactor pays for itself in less than seven years.

So, what starts as a clean energy plant quickly becomes something much more: a revenue-generating, multi-output powerhouse that supports both the electric grid and lifesaving medicine.

Selling water and brine byproducts: Imagine we build a large reverse osmosis water system that can clean about 100,000 cubic meters of seawater every day. That's roughly 26 million gallons daily, enough to serve a large city. This system uses electricity to push seawater through filters and heat to help handle the leftover brine. Altogether, it would need about 15 megawatts of electricity and another 20 megawatts of heat (what engineers call "low-grade" heat).

To build everything—the filters, pumps, and special equipment to dry out the salty waste without dumping it into the ocean—it would cost around $150 million up front. But here's where thorium comes in: by using the waste heat from a thorium reactor, the system becomes much more efficient. That reduces the extra cost of treating the leftover brine to just 25 cents per cubic meter of water.

Now, let's talk about money. If the water is sold at a typical price of $1.20 per cubic meter, the plant could earn about $36 million each year. Running the system (replacing filters, maintaining pumps, and so on) costs around $13 million a year. That leaves $23 million in profit every year. At that rate, the whole system would pay for itself in under six years.

And that's not all. The dried-out salts left over from desalination aren't waste. They can be sold for industrial uses, like making chlorine or de-icing roads in the winter. That could bring in another $4 million a year. So not only does the plant clean water efficiently, but it also turns its waste into a product people will pay for. In short: clean water, good profits, and no waste dumped in the sea. That's the power of pairing reverse osmosis with thorium energy.

Adding sales from synthetic fuel production: Now let's take things a step further and add synthetic fuel production to the mix. Imagine attaching a fuel-making module to the same thorium reactor we just talked about. According to a 2025 financial model by nuclear energy advocate Robert Hargraves, the cost to set that up, including solid-oxide electrolysis, Fischer–Tropic fuel synthesis, and storage tanks, would be around $180 million.[35]

Using the extra heat and about 15 megawatts of surplus electricity from the reactor, the system could produce roughly 32,000 barrels of synthetic diesel every year. This isn't just any diesel: it's sulfur-free, clean-burning e-fuel. At a conservative wholesale price of $140 per barrel, the revenue adds up fast. After covering the energy costs, there's about $4.5 million in profit.

But there's more. The electrolysis process also creates oxygen as a byproduct: something that can be sold to nearby green steel plants or even wastewater treatment facilities. That adds another $2 million a year. Altogether, this fuel module brings in about $5 million in additional annual revenue.

Now zoom out and look at the full site: power, water, medical isotopes, and synthetic fuel—all running from the same thorium core. With total capital costs of around $530 million, this setup could generate $58 million in annual revenue. That's a solid return, somewhere in the mid-teens as a percentage, even before counting carbon credits or policy incentives.

In short, this goes beyond clean energy. It's a next-generation industrial platform; one that produces electricity, water, fuel, and medicine, all with minimal waste and strong economics.

Stacking Revenues from Co-Production

Layer	Added capital (first-of-a-kind)	Annual net cash flow	Payback contribution
Base electricity regional grid + AI cluster	$200 million (DOE ARDP benchmark $4,000 kW*)	$24 million (8,000 h × 50 MW × $70 MWh minus $4 MWh O&M)	≈ 9 years
Medical-isotope loop (Bi-213, I-131)	+ $40 million (hot cell, shielding)	+ $6–7 million (900 Ci y^{-1} × $8 000 mCi less operating cost)	Payback < 7 years
Desalination + ZLD 100 000 $m^3 d^{-1}$	+ $150 million (RO, pumps, crystallizer)	+ $23 million (water at $1.20 m^3 minus $0.45 m^3 cost) + $4 million salt sales	Payback < 6 years
"Seafinery" e-diesel & hydrogen	+ $180 million (SOEC, Fischer–Tropsch)	+ $5 million (32,000 bbl y^{-1} at $140 bbl plus O_2 sales)	Site IRR mid-teens

* US Advanced Reactor Demonstration Program cost basis (2024).

So why does stacking all these revenue streams work so well?

It comes down to three key reasons.

First, diversification. Electricity prices, water rates, medical isotope demand, and synthetic fuel markets don't all move together. When one line of business dips, another usually holds steady or even rises. That helps cushion the ups and downs.

Second, shared infrastructure. We only need one reactor. One operating team. One security crew. Every new output—whether it's water, fuel, or isotopes—gets to reuse the same fixed costs. That makes each add-on more affordable and more profitable.

Third, policy tailwinds. Today's clean energy rules actually reward this kind of setup. Carbon fees, low-emission fuel standards, and even drought resilience grants can all boost the value of these non-electric products.

Energy advocate Robert Hargraves calls this "the Swiss army knife model of new nuclear." And the International Atomic Energy Agency agrees. In every regional scenario they've modeled, multi-product reactors outperform single-purpose plants on return and payback time.

So, if you're an investor or utility used to razor-thin electricity margins, the thorium stack opens up a much clearer path to profits. And if you're a policymaker, this kind of system gives you a powerful toolkit: solving for clean power, clean water, clean fuels, and cancer treatments, all in one footprint.

Where the Money Comes From: New Ways to Fund Thorium

Thorium is attracting a new class of investors. Green bonds, which are financial tools designed to support environmentally friendly projects, are being used to fund nuclear innovation. Venture capitalists, who typically fund tech startups, also consider thorium a long-term play.

Early-stage backing from philanthropic climate investors, especially for co-located water and food projects, could unlock mainstream infrastructure funding within a decade.

International development banks are also poised to support thorium-powered desalination and microgrid pilots in acute water-energy poverty regions.

The 2023 Climate Bonds Initiative listed advanced nuclear among its "Emerging Green Infrastructure" categories, citing modular molten salt reactors as promising clean water-energy systems.[36]

Investment in thorium technology is required for funding research, engineering, and testing to make it scalable. Investors recognize its potential in the future energy market. Additionally, governments support thorium projects with tax incentives and grants, boosting momentum and confidence. Public funding encourages private sector involvement in building these technologies by reducing risks in the early stages.

Creating a Thriving Market

Smart financial incentives can make a big difference. These include:

Incentive type	Purpose
Tax credits	Reduce upfront capital and ongoing operating costs of thorium reactors
Feed-in tariffs	Ensure a predictable price for each kilowatt-hour of thorium-generated electricity
Innovation grants	Bridge the funding gap between laboratory prototype and first commercial unit
Public risk-sharing	Let governments co-fund or underwrite early projects, lowering financial exposure for private firms

Incentives give entrepreneurs and investors the confidence to commit to thorium development. It also signals to the market that it's a serious player in the global energy transition.

Modeling from the *Journal of Cleaner Production*

As fossil fuel supplies run low, their price will climb, and nuclear alternatives will become increasingly attractive. A techno-economic analysis published in 2018 in the *Journal of Cleaner Production* evaluated nuclear-powered desalination using advanced reactors. Thorium-based systems could become economically superior within 20–30 years, especially considering carbon pricing or energy security. While the study is not exclusively focused on thorium, it includes thorium reactors as part of its future-oriented modeling.[37]

Private Sector and Venture Interest

Firms like Flibe Energy are attracting early-stage private investment and interest from the Department of Energy in thorium reactors for co-generation uses like desalination, as published in their 2023 study.[38] Rising financial backing signals market confidence in the long-term ROI of such projects.

The World Economic Forum and the International Atomic Energy Agency have identified advanced nuclear technology, including thorium reactors, as a critical part of the clean energy transition. The IAEA notes that thorium's resistance to proliferation, lower waste output, and abundant global supply make it especially attractive to countries with established nuclear programs, like Canada, India, and China, as a means to extend those benefits into water security and energy independence.

Startup Momentum: Small Companies Leading the Thorium Revival

While government labs and legacy energy firms have traditionally dominated nuclear development, a new wave of small businesses and startups is accelerating innovation in molten salt reactor technology, including thorium-based systems. These agile companies are combining advanced engineering with creative business models to reimagine nuclear power for the 21st century.

Seaborg Technologies, based in Copenhagen, Denmark, is among the most visible. With over $28 million in funding and a team of more than 100 engineers and developers, Seaborg is building a Compact Molten Salt Reactor (CMSR). Designed for maritime deployment, Seaborg's CMSR runs on molten fluoride salts and is tailored for modular construction and rapid grid integration. Their goal is to deliver plug-and-play reactor barges that can power coastal cities, desalination hubs, or hydrogen plants without the need for large-scale infrastructure.

Core Power, in the United Kingdom, is charting a similar course, but focused squarely on the marine shipping sector. With $14.3 million raised and a lean, specialized team, Core Power is developing a marine molten salt reactor (m-MSR) intended to revolutionize long-haul cargo transport. By replacing bunker fuel with zero-emission nuclear propulsion, their technology could radically reduce carbon output in one of the world's dirtiest industries: international shipping.

Thorizon, in The Netherlands, is a venture-backed startup with over $45 million in funding. Thorizon is advancing molten salt reactor designs with a focus on recyclability and waste reduction. Their approach includes converting existing nuclear waste into usable fuel, effectively turning a liability into an energy asset. By targeting industrial markets and emphasizing circular fuel cycles, Thorizon represents the intersection of clean energy and sustainable waste management.

Gen4 Energy, based in Santa Fe, New Mexico, is working on mini reactors for remote, off-grid, or defense-related applications. With funding exceeding $16 million, Gen4 aims to deliver safe, transportable nuclear power units that can operate in isolated or rugged environments. Their microreactor concepts align with military logistics, disaster recovery zones, and small communities with unreliable grid access.

What unites these startups is their shared conviction that the future of nuclear must be modular, scalable, and commercially viable. Many are exploring or advancing thorium-compatible molten salt systems, including small businesses already mentioned earlier such as **Flibe Energy** in Alabama and **Copenhagen Atomics** in Denmark. These ventures reflect a growing movement of innovators reshaping nuclear power into a dynamic source for clean energy, clean water, clean shipping, and even medical isotope production. In this new landscape, agility, specialization, and innovative business models are as crucial as reactor technology.

Thorium as the New Industrial Gold Rush?

Countries that invest in thorium early can capture entire industrial value chains from reactor manufacturing to synthetic fuels, medical isotopes, and AI-powered energy management systems.

Investors should think seriously about this. Just as China dominated solar panels and South Korea built global shipyards, thorium supply chain leadership offers decades of economic advantage.

The World Nuclear Association projected in 2023 that the global market for SMRs and advanced nuclear services could exceed $500 billion annually by 2050. This may be the optimal time for entry to minimize risks and maximize returns.

Just as a fire cannot burn without a catalyst sustaining the reaction, so too would all these innovations fail if not for the passion and determination of the innovators who persist. Next we'll look at the people, the partnerships, and the countries who are carrying the torch for thorium innovation.

The Catalysts: Who are the Innovators?

Innovation Through Collaboration

The Massachusetts Institute of Technology (MIT) is an advanced nuclear energy research hub. Its Department of Nuclear Science and Engineering is one of the most active in the world, producing breakthrough work in areas like reactor safety, fuel cycle innovation, and MSR modeling. MIT primarily conducts experiments with uranium-based fuels in its reactors. However, the institute also supports research on thorium-based fuel cycles. This work happens mainly through its Nuclear Energy Futures Lab and partnerships with groups like Flibe Energy. MIT researchers have published studies on thorium-fueled MSRs, evaluating their potential for safe, long-term electricity generation with less waste. The institute has also organized panels and workshops to discuss thorium fuel cycles, especially focusing on their use in developing countries.[39]

At CERN, the Conseil Européen pour la Recherche Nucléaire, or in English, the European Organization for Nuclear Research, which is best known for particle physics, researchers are studying how thorium can be used in accelerator-driven systems (ADS) for subcritical reactors. One such project is the MYRRHA program, which is co-funded by the European Commission and Belgium's SCK CEN.

Why does this research matter? A subcritical reactor is analogous to a fire that cannot stay lit on its own because it needs a steady spark to keep burning. In an accelerator-driven system, that spark comes from a machine that fires tiny particles at a metal target, knocking loose extra neutrons. Those neutrons let the thorium give off heat, but only while the machine is running. Turn the machine off, and the reaction dies down almost instantly, adding another built-in layer of safety.

SCK CEN, the Studiecentrum voor Kernenergie – Centre d'Étude de l'Énergie Nucléaire, or in English, the Belgian Nuclear Research Centre, is conducting the MYRRHA program to look at using thorium alongside proton accelerators to create safe, low-waste energy. CERN has also supported the basic physics involved in these processes of spallation and transmutation.

This research matters because it could show us how to make nuclear power both safer and cleaner. Continuing the fire analogy, the MYRRHA project plans to run a "reactor that needs a spark," so it can never run out of control, and it burns fuel that leaves only a small amount of long-lived waste. Thorium is cheap and widely available, and the accelerator adds an extra shut-off switch: flip it off and the reaction stops. CERN's work on the particle physics behind that spark, how high-speed protons dislodge neutrons from heavy metals and transform one element into another, gives MYRRHA the scientific know-how it needs to turn the idea into a working power source.[40]

MIT and CERN are two key players in nuclear innovation. MIT focuses on designing and building practical nuclear reactors, while CERN studies basic nuclear interactions. Both areas are important for making thorium energy a reality.

Bringing People Together: The Power of Partnerships

The private sector can't do this alone, and neither can governments. Through partnerships, research labs, energy companies, regulators, and investors have built the most successful thorium projects. These collaborations allow for shared risk, more innovative design, and faster progress. Although not fueled by thorium, the United Arab Emirates' (UAE) nuclear energy rollout is a strong example of how public-private partnerships can work.

The UAE's Nuclear Energy Program: A Successful Example of Public-Private Partnership

The UAE showcases how public-private partnerships (PPPs) can drive major energy projects effectively. The Barakah Nuclear Power Plant is a key example, costing $24.4 billion and relying on global collaboration between government and private firms.

The Emirates Nuclear Energy Corporation (ENEC) led the effort, providing the vision, regulations, and funding necessary for the project. Instead of developing the technology independently, ENEC partnered with Korea Electric Power Corporation (KEPCO), which contributed expertise and knowledge from South Korea's successful nuclear industry.

The Barakah project is the first nuclear power plant in the Arab world, set to generate 5.6 gigawatts and supply up to 25% of the UAE's electricity. In 2023, three of its four reactors became operational. It is expected to reduce CO_2 emissions by over 21 million tons annually, offering environmental and economic advantages.

The partnership created over 2,500 direct jobs during peak construction and continues offering skilled operations and maintenance positions. KEPCO and its subsidiaries trained local Emirati engineers, building long-term local expertise.

This PPP helped the UAE quickly advance its nuclear program by using global private-sector talent while controlling safety and policy.

The Barakah project shared risks between the government and the private sector. The UAE government provided long-term funding and clear regulations through the Federal Authority for Nuclear Regulation (FANR). Meanwhile, KEPCO and other contractors contributed technical skills and took on specific performance tasks. This cooperation helped keep the project on track and ensured safety and transparency.

The World Nuclear Association and World Bank have recognized Barakah as a model for other countries looking to develop nuclear programs through international teamwork and public-private investment. It shows emerging nuclear nations can build advanced infrastructure with good governance, expertise, and shared goals.

Worldwide Case Studies in Thorium Economic Transformation

Thorium reactors can help countries become more self-sufficient, reducing their reliance on energy imports. By investing in thorium, countries can protect themselves from rising fuel prices, supply disruptions, and geopolitical conflicts that affect energy.

Control over energy sources leads to healthier economies, better trade deals, and reduced inflation risks, especially as fossil fuel markets experience more ups and downs.

A 2022 World Bank report shows that countries with their own local energy supplies have 25% less GDP volatility than those that rely on energy imports.

Norway: In Norway, the path to thorium power is unfolding through smart collaboration. The country's Institute for Energy Technology, or IFE, along with Thor Energy, a company based in Oslo, is working closely with Westinghouse, the U.S. Department of Energy, and several European utilities. Together, they're testing a next-generation fuel blend that combines thorium with MOX, or mixed oxide fuel.[41] The tests are being carried out at the OECD's Halden research reactor - one of the most respected facilities of its kind.

What makes this project so important is that it's generating the first modern performance data on commercial-grade thorium fuel.[42] That's a major step forward, because until now, most of what we've known about thorium has come from decades-old experiments.

But Norway isn't doing this alone. The research results are being shared with labs in the Netherlands, France, and the Czech Republic through two European Union–funded programs called SAMOFAR and SAMOSAFER. These programs are helping scientists build better safety models for molten salt reactors by pooling expertise from across the continent.

At the same time, IFE is also working with Canadian Nuclear Laboratories to qualify thorium fuel for future regulatory approval. It's a smart strategy: spread the research across multiple countries, share the data, and reduce the risks and costs of development for everyone involved. Norway's approach shows how international partnerships can accelerate progress, especially with a technology as promising and as complex as thorium.

Canada: Canada's CANDU heavy-water reactors demonstrate a strong return on investment for nuclear energy. Their reliable design and affordable upgrades have kept most reactors operating at over 90% capacity for many years, which ensures steady income for utilities.

One of the advantages of CANDU reactors is that their lattice accepts many fuel types and can be refueled while online, so a CANDU core could switch to thorium simply by changing bundle geometry and control software. A 2015 study found that retrofitting these reactors to use thorium would reduce fuel costs and produce less waste, all while using the existing reactor structure. This upgrade offers high payback potential.[43]

Canadian Nuclear Laboratories and AECL have been looking into using small modular reactors powered by thorium and CANDU systems for desalination, particularly in northern areas. Their analysis from 2017 indicates that these technologies could be cheaper than using diesel for water and electricity within the next 15 to 20 years. This cost savings would primarily come from reduced carbon fees and lower transportation costs for fuel. The studies suggest that Canada is in a strong position to test and develop commercial thorium energy projects.[44]

Indonesia: Indonesia is taking major steps toward building its first advanced nuclear power plant, and while thorium is part of the long-term vision, it's not the fuel of choice just yet. The country has partnered with ThorCon, a company developing a shipyard-built molten salt reactor. Their current design, known as the ThorCon 500, is based on uranium fuel, specifically low-enriched uranium, and not thorium. This choice was intentional, aimed at simplifying licensing and deployment by using a fuel type that's already well understood by regulators.

The plan is to construct the 500-megawatt reactor modules in a shipyard, using the same techniques used to build large ocean-going vessels. Once completed, each module is floated to shore and installed at a prepared site. This drastically cuts down on construction time and cost, which is especially important for nations like Indonesia that face both power shortages and rising electricity demand.

The long-term vision, however, includes thorium. ThorCon has stated that its reactor architecture is thorium-capable. That means once the uranium-fueled version is proven and operational, future models could transition to using thorium fuel, especially as regulatory frameworks evolve and thorium supply chains mature.

For Indonesia, this partnership represents more than just a new power plant. It's a strategic move toward energy independence, grid stability, and climate resilience. And if successful, it could serve as a template for other tropical and island nations struggling with chronic energy shortfalls and limited access to freshwater. While the first version runs on uranium, the reactor design leaves the door wide open for a thorium-powered future.

India: India is home to the largest known reserves of thorium in the world, but it isn't using thorium as a primary fuel just yet. Currently, most of India's nuclear power comes from uranium-fueled reactors, including pressurized heavy water reactors and a few newer designs that still rely on imported uranium. But India has long recognized thorium's potential, especially because it offers a path to energy independence using resources found within its own borders.

That's why India's Department of Atomic Energy has been steadily developing the Advanced Heavy Water Reactor, or AHWR. This reactor is specifically designed to run on thorium once it's paired with a small amount of fissile material to start the reaction. It's a major part of India's long-term three-stage nuclear program, which aims to shift from uranium to a self-sustaining thorium fuel cycle over time.

According to a 2023 report published by the World Nuclear Association, the AHWR program projects that, once scaled, thorium-fueled reactors could reduce electricity generation costs by as much as 30% compared to building new coal plants. The benefits don't stop at clean power. When thorium reactors are paired with desalination, they may also beat fossil-fueled plants on cost, especially when we factor in long-term savings on fuel imports and carbon emissions.[45]

For India, the payoff could be huge: not just affordable clean energy and abundant water, but true energy independence for one of the world's fastest-growing economies. While thorium isn't powering India's grid yet, the country is investing seriously in getting there and building the infrastructure to make it happen.

China: In 2021, China's National Nuclear Corporation launched an experimental molten salt thorium reactor in Wuwei, Gansu Province, the first approved for full operation. This reactor is part of a broader $3.3 billion program to demonstrate the viability of thorium fuel cycles in the SMR form.[46]

A 2023 Institute of Nuclear and New Energy Technology report at Tsinghua University estimates that thorium-fueled MSRs could break even in under 18 years, well ahead of typical nuclear investments, with ROI exceeding 8% annually when paired with desalination or industrial heat applications. These figures are especially notable given that early-stage nuclear technologies typically see break-even points of 25 years or longer.[47]

According to analysis from the Chinese Academy of Sciences, these reactors are expected to produce electricity at a levelized cost of energy (LCOE) between $0.045 and $0.065 per kilowatt-hour. That is on par with or below the cost of coal-fired plants, particularly when externalities like carbon pricing or health impacts are considered. When desalination capabilities are factored in, the value proposition improves further.

China's national strategic plan includes the potential to integrate these reactors with coastal desalination facilities, especially in water-scarce northern provinces. China projects that large-scale thorium use will lower uranium and fossil-fuel imports by over $1 billion a year within a decade, savings it plans to reinvest in modular plants across the country by 2035.[48]

These countries offer key lessons: collaborate to accelerate progress (Norway), scale for high efficiency (Canada), use industries like shipyards to turn out scalable reactors (Indonesia), build where resources exist (India), and take bold steps with prototype designs (China). Each illustrates how thorium is getting factored into national energy strategies.

Why Talk About Risk Now?

Even the most elegant technology stalls if decision-makers can't picture its downside and the plan to manage it. The matrix below distills the top technical, regulatory, financial, and social hurdles facing thorium-fueled SMRs, then pairs each with proven or emerging mitigation levers and clear ownership lines. Think of it as a quick-look playbook: a snapshot of where the roadblocks lie and who must move first to clear them.[49-56]

Risks and Mitigation

Risk Theme	Specific Concern	Mitigation Lever	Who Leads?
Technical maturity	Salt corrosion & materials embrittlement	Multiyear coupon testing in prototypic FLiBe loops; accelerated irradiation in HFIR/ATR	National labs, materials vendors
Regulatory uncertainty	No MSR-specific licensing framework	Early, pre-application engagement; use of risk-informed, performance-based rules piloted with NRC Part 53 draft	Reactor developers + regulators
Supply-chain gaps	Specialty nickel alloys, Li-7 purification, HALEU availability	Joint-venture alloy foundries; DOE down-blend program; incentivized Li-7 production line	Private–public partnerships
Capital overruns	First-of-a-kind learning curve	Factory modularization; milestone-based finance tranches; EPC wrap guarantees	Developers, EPC firms, lenders
Public perception & siting	Nuclear stigma, local opposition	Community benefits agreements; transparent data sharing; micro-grant STEM programs	Utilities, municipalities
Spent-salt management	On-site interim storage & eventual disposition route	Integrate on-line fission-product extraction; commit to deep-borehole or partition-&-transmute pathway	Industry consortium + DOE

Summary: Innovation at the Crossroads

Thorium innovation has the potential to drive considerable advancements across various sectors. It can provide solutions like water in deserts, clean ships at sea, food in cold climates, energy for artificial intelligence, and power for underserved communities. To realize this transformation, we need innovators to address challenges, investors to fund new technologies, and advocates to spread the word.

We've considered its impact on electricity, water security, and food production. Next, we will look at how thorium reactors unlock resilient, integrated systems for water purification, electricity generation, and climate adaptation.

1. US NAVY, FIRE FIGHTING FUNDAMENTALS, NRTC 14057 (PENSACOLA, FL: NAVAL EDUCATION AND TRAINING COMMAND, 1993), CHAP. 4.

2. INTERNATIONAL ENERGY AGENCY, SMALL MODULAR REACTORS: OUTLOOK AND ROADMAP (PARIS: IEA, 2022), 11–13.

3. US DEPARTMENT OF ENERGY, "MICROREACTOR PROGRAM OVERVIEW," OFFICE OF NUCLEAR ENERGY FACTSHEET, 2021.

4. IDAHO NATIONAL LABORATORY, MOLTEN SALT REACTOR TECHNOLOGY READINESS (INL/EXT 20 59023, 2020), 6.

5. US NUCLEAR REGULATORY COMMISSION, "NUSCALE VOYGR SMR COLA STATUS," DOCKET 99902078, DECEMBER 2023.

6. WORLD NUCLEAR ASSOCIATION, "NUSCALE SMALL MODULAR REACTOR," UPDATED 2023.

7. US DEPARTMENT OF ENERGY, "NUSCALE COST SHARE TERMINATION NOTICE," PRESS RELEASE, 2023.

8. OAK RIDGE NATIONAL LABORATORY, CHEMICAL STABILITY OF FLIBE SALT COOLANT (ORNL/TM 2021 1422, 2021).

9. FLIBE ENERGY, "WHY MOLTEN SALT REACTORS," TECHNICAL BRIEF, 2022.

10. KAIROS POWER, "TECHNOLOGY OVERVIEW," COMPANY WHITE PAPER, 2023.

11. WORLD NUCLEAR ASSOCIATION, ADVANCED NUCLEAR POWER REACTORS – SMR SECTION, LAST UPDATED APRIL 2024.

12. CHINA NATIONAL NUCLEAR CORPORATION, "THORIUM MSR DEMONSTRATION IN GANSU PROVINCE," PRESS RELEASE, SEPTEMBER 2021.

13. INTERNATIONAL ENERGY AGENCY, DATA CENTRES AND DATA TRANSMISSION NETWORKS (PARIS: IEA, 2023), 8–9.

14. US DEPARTMENT OF ENERGY, STRATEGY FOR THE MANAGEMENT AND DISPOSAL OF USED NUCLEAR FUEL AND HIGH LEVEL RADIOACTIVE WASTE (JANUARY 2021)

15. TERRELL, JEFF. "COPENHAGEN ATOMICS: THE DANISH STARTUP BUILDING MASS-PRODUCED THORIUM REACTORS." MEDIUM, MARCH 5, 2023. HTTPS://MEDIUM.COM/@TERRELLJEFF/COPENHAGEN-ATOMICS-THORIUM-VISION

16. WORLD NUCLEAR NEWS. "COPENHAGEN ATOMICS SIGNS MOU FOR AMMONIA PRODUCTION IN INDONESIA." WNN, APRIL 2024. HTTPS://WORLD-NUCLEAR-NEWS.ORG

17. INTERNATIONAL ENERGY AGENCY. ELECTRICITY 2024: ANALYSIS AND FORECAST TO 2026. PARIS: IEA, JANUARY 2024. HTTPS://WWW.IEA.ORG/REPORTS/ELECTRICITY-2024.

18. Vincent, roger. "amazon teams up with nuclear power plant to fuel its data centers." Los angeles times, january 22, 2024. Https://www.latimes.com/business/story/2024-01-22/amazon-nuclear-data-center-pennsylvania.

19. Federal energy regulatory commission. Order denying request for increased interconnection capacity at susquehanna nuclear facility. Docket no. Er24-151, february 2024. Https://www.ferc.gov.

20. U.s. department of energy. Doe approves $1.5 billion loan to restart palisades nuclear plant. Office of nuclear energy press release, march 2024. Https://www.energy.gov/ne/articles/doe-approves-loan-palisades-restart.

21. Sverdlik, yevgeniy. "data centers in northern virginia face seven-year wait for power." Data center knowledge, november 14, 2023. Https://www.datacenterknowledge.com/power-and-cooling/data-centers-northern-virginia-face-seven-year-wait-power.

22. Argonne national laboratory. Ai-driven tools enhance nuclear reactor design and operation. U.s. department of energy office of science highlight, october 2023. Https://www.anl.gov/article/ai-driven-tools-enhance-nuclear-reactor-design-and-operation.

23. International atomic energy agency. Artificial intelligence for nuclear technology and applications, iaea-tecdoc-2024-02. Vienna: iaea, 2024.

24. Us department of energy. "digital twin technology in nuclear reactor operation." Office of nuclear energy, october 2023. Https://www.energy.gov/ne/digital-twin-nuclear-reactors

25. International atomic energy agency. Industrial applications of nuclear energy: a path to decarbonization and development. Iaea-tecdoc-1999, 2022.

26. Cryogenic society of america. "energy use in cryogenics." Cold facts, spring 2023. Https://www.cryogenicsociety.org

27. Us department of energy. Energy requirements for high-tech cooling systems, office of energy efficiency and renewable energy, 2024.

28. Terrapower, "medical isotopes," terrapower isotopes, accessed june 2025, https://www.terrapower.com/what-we-do/isotopes/.

29. Us department of energy, advanced nuclear isotope production strategy report, april 2024, https://www.energy.gov/ne/advanced-nuclear-isotope-strategy.

30. INTERNATIONAL ENERGY AGENCY, SMALL MODULAR REACTORS: OUTLOOK AND ROADMAP (PARIS: IEA, 2022), 11–13.

31. INTERNATIONAL ATOMIC ENERGY AGENCY, "COST–BENEFIT ANALYSIS OF NUCLEAR DESALINATION," IAEA-TECDOC-2000 (VIENNA: IAEA, 2022), 17.

32. WORLD BANK, HIGH AND DRY: CLIMATE CHANGE, WATER, AND THE ECONOMY (WASHINGTON, DC: WORLD BANK, 2016), XIII.

33. HARGRAVES, ROBERT. THORIUM: ENERGY CHEAPER THAN COAL (WHITE RIVER JUNCTION: CHELSEA GREEN PUBLISHING, 2012).

34. US DEPARTMENT OF ENERGY, ADVANCED REACTOR DEMONSTRATION PROGRAM COST BASIS REPORT (WASHINGTON, DC: DOE, 2024), 12–13.

35. ROBERT HARGRAVES, NEW NUCLEAR IS HOT IN 2025 (WHITE RIVER JUNCTION, VT: HARGRAVES, 2025), 61–65.

36. CLIMATE BONDS INITIATIVE, ADVANCED NUCLEAR: EMERGING GREEN INFRASTRUCTURE (LONDON: CBI, JULY 2023).

37. J. WU, Y. LI, AND Z. ZHANG, "TECHNO-ECONOMIC ASSESSMENT OF NUCLEAR-POWERED DESALINATION," JOURNAL OF CLEANER PRODUCTION 172 (2018): 3683–94.

38. FLIBE ENERGY, MOLTEN SALT REACTOR TECHNOLOGY OVERVIEW (HUNTSVILLE, AL: FLIBE, 2023).

39. MIT NUCLEAR ENERGY FUTURES LAB, ASSESSMENT OF THORIUM-FUELED MOLTEN-SALT REACTORS FOR EMERGING ECONOMIES (CAMBRIDGE, MA: MIT, 2022).

40. SCK CEN, "MYRRHA PROJECT OVERVIEW," ACCESSED MARCH 2025, HTTPS://MYRRHA.BE.

41. INSTITUTE FOR ENERGY TECHNOLOGY, "THORIUM-MOX FUEL QUALIFICATION IN HALDEN," PROJECT FACT SHEET (KJELLER, 2023).

42. WESTINGHOUSE ELECTRIC CO. AND THOR ENERGY, "THORIUM–MOX FUEL IRRADIATION DATA PACKAGE," PRESENTATION TO OECD-NEA HALDEN PROJECT, 2024.

43. S. GOURISHANKAR AND M. LEONARD, "THORIUM FUEL OPTION FOR OPERATING CANDU REACTORS," NUCLEAR ENGINEERING AND DESIGN 295 (2015): 510–19.

44. CANADIAN NUCLEAR LABORATORIES, SMALL MODULAR REACTORS FOR NORTHERN DESALINATION (CHALK RIVER, ON: CNL, 2017).

45. WORLD NUCLEAR ASSOCIATION, "ADVANCED HEAVY WATER REACTOR (AHWR)," LAST UPDATED OCTOBER 2023.

46. CHINESE ACADEMY OF SCIENCES, "THORIUM MOLTEN SALT REACTOR EXPERIMENTAL RESEARCH PROGRAM," 2022.

47. INSTITUTE OF NUCLEAR AND NEW ENERGY TECHNOLOGY (TSINGHUA UNIVERSITY), ECONOMIC PROSPECTS FOR THORIUM MOLTEN-SALT REACTORS, INET-R-2023-06 (BEIJING, 2023).

48. CHINESE ACADEMY OF SCIENCES, "LEVELISED COST OF ENERGY FOR ADVANCED REACTORS," POLICY BRIEF (BEIJING, 2023).

49. INTERNATIONAL ATOMIC ENERGY AGENCY, PROLIFERATION RESISTANCE IN ADVANCED REACTORS, IAEA TECDOC 2045 (VIENNA: IAEA, 2024).

50. IDAHO NATIONAL LABORATORY, FLUORIDE SALT HANDLING GUIDELINES (INL/EXT 21 60011, 2021).

51. NUCLEAR ENERGY INSTITUTE, "BEST PRACTICES FOR ADVANCED REACTOR LICENSING ENGAGEMENT," NEI WH US DEPARTMENT OF ENERGY, ADVANCED REACTOR DEMONSTRATION PROGRAM COST BASIS REPORT (WASHINGTON, DC: DOE, 2024), 12–13.

52. INTERNATIONAL ATOMIC ENERGY AGENCY, REGULATORY EXPERIENCES WITH LICENSING SMALL MODULAR REACTORS (VIENNA: IAEA, 2023).

53. OAK RIDGE NATIONAL LABORATORY, CHEMICAL STABILITY OF FLIBE SALT COOLANT (ORNL/TM-2021-1422, 2021).

54. ARGONNE NATIONAL LABORATORY, AI-ASSISTED MOLTEN-SALT REACTOR CHEMISTRY CONTROL (ANL/ES-23/45, 2023).

55. WORLD BANK, HIGH AND DRY: CLIMATE CHANGE, WATER, AND THE ECONOMY (WASHINGTON, DC: WORLD BANK, 2016).

56. CLIMATE BONDS INITIATIVE, ADVANCED NUCLEAR: EMERGING GREEN INFRASTRUCTURE (LONDON: CBI, 2023).ITEPAPER, 2023.

Enjoying the Journey So Far?

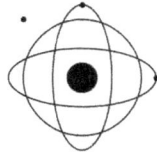

Want to help others discover thorium's possibilities?
If you're finding value in **Thorium-Powered Abundance**, I'd be
truly grateful if you took a moment to leave a quick review at the site
where you
attained this book.

Please write a review.

Your feedback helps others find this book, and supports the mission to
bring abundant clean energy and water to the forefront of public
conversation.

Even a sentence or two makes a big difference.

Thank you,

Michael Lee Anderson
Author, Engineer, and Thorium Optimist

Part Three

Part III: Fixing the Grid, Feeding the World

Chapter 6

Power Where It's Needed: Water and Fuel

The Water-Energy Nexus

The water-energy nexus refers to the connection between water resources and energy production, which is essential for modern society. Traditional systems often struggle because they consume significant amounts of water for energy production or require substantial energy to purify water. This cycle puts pressure on available resources, especially in water-scarce areas. To address these issues, we need integrated solutions that optimize water and energy use, reduce waste, and enhance resilience to climate change.

Water-Energy-Food Nexus Powered by Thorium

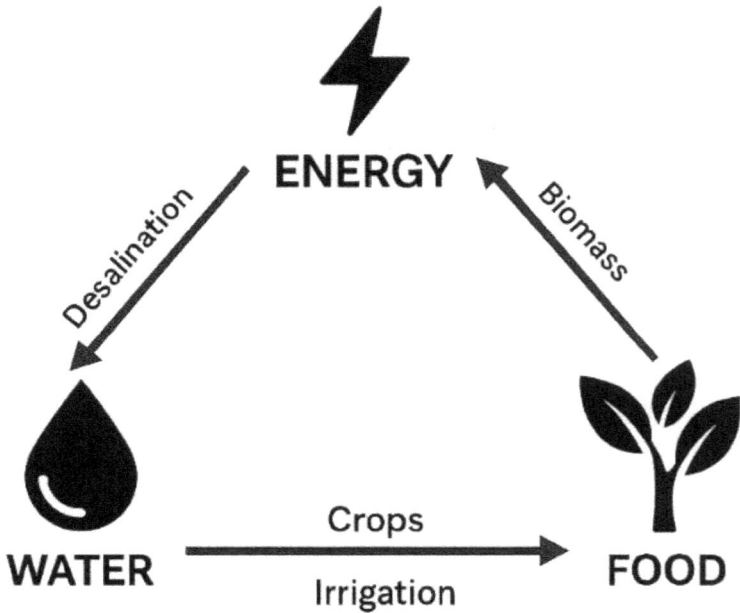

Thorium-powered facilities can help tackle water scarcity and energy demand while minimizing environmental impact. These reactors can use waste heat to generate desalinated water alongside electricity. We can effectively manage resources in regions facing shortages by pairing thorium reactors with advanced water management technologies and smart grids. This approach decreases our dependence on fossil fuels and ensures a stable supply of clean water and electricity for growing populations.

Addressing Thirst with Real Reactors, Real Results

Current projects in water-stressed areas demonstrate how thorium can effectively manage resources and support sustainability objectives, with recent pilot efforts proving its potential in practice.

India's Department of Atomic Energy has operated a combined water and energy plant in Kalpakkam, producing 6,300 cubic meters of clean water per day since 2005. The plant uses a heavy-water reactor to power both reverse osmosis and multi-effect water purification systems. Simultaneously, scientists there are running long-term experiments with molten salts containing thorium. They're testing how these hot, reactive fluids affect the corrosion of different metals over time. By running these experiments alongside the operating desalination plant, scientists gain real-world performance data, essential for choosing the right materials to safely contain thorium fuel in future reactors.

It's a foundational step in India's three-stage effort to design and eventually deploy thorium-fueled molten salt reactors, using their vast domestic thorium reserves.[1]

India's Three-Stage Nuclear Power Program

INDIA'S THREE-STAGE NUCLEAR POWER PROGRAM

STAGE 1
Heavy Water
Reactors

STAGE 2
Fast Breeder
Reactors

STAGE 3
Advanced Thorium
Reactors

In Indonesia, ThorCon and the state utility PLN signed a 2024 framework agreement for a 500-MW thorium molten-salt vessel off Bangka Island that would send one-third of its output to a co-located 100,000 cubic meter per-day RO unit serving water-stressed coastal villages.[2]

The United Arab Emirates' Emirates Nuclear Energy Corporation, Khalifa University, and Flibe Energy announced a 2023 study on siting a thorium MSR next to an existing seawater-intake channel at Ruwais, to produce 80,000 cubic meters of clean, waste-free water daily while reducing the region's dependence on gas-powered plants.[3]

These projects provide regulators and investors with the data needed to assess how thorium can address power and water shortages in dry regions. But water isn't the only resource in seawater. With thorium's continuous power and high heat output, the same site that produces clean water could also extract the carbon and hydrogen needed to make synthetic fuels, turning coastal desalination hubs into energy manufacturing zones.

Responsive Power: Thorium for a Dynamic Grid

Powering water systems is just one application of thorium. With the growing share of renewables like solar and wind, power systems need to be dynamic and responsive. Thorium reactors, especially molten salt designs, are ideal for this due to their steady heat, low pressure, and high efficiency.

As electric grids increasingly rely on variable renewable energy sources, there is a rising demand for clean, dispatchable energy that can be adjusted based on needs. Unlike traditional coal and nuclear plants that operate at constant output, thorium molten salt reactors can easily modulate their heat output. They can reduce turbine load during high renewable production by sending excess heat to thermal storage. When demand increases or renewable output decreases, this stored thermal energy can quickly generate electricity, maintaining grid balance without the need for carbon-heavy peaker plants.[4]

Conventional light-water reactors are complex and slow to adjust output, making them less flexible than molten salt reactors. Many proposed uranium-based small modular reactors (SMRs) also lack the real-time flexibility of thorium systems. Thorium MSRs can operate dynamically while ensuring high efficiency and safety, making them pivotal for the transition to cleaner energy systems.

The US Department of Energy has modeled scenarios where molten salt reactors adjust their output according to solar peaks and night-time consumption, showing significant improvements in grid stability. Integrating these systems into modern transmission networks will require regulatory adjustments, but the capacity of thorium reactors to enhance grid resilience is evident.[5]

As utilities build hybrid systems that combine renewables, energy storage, and flexible power generation, thorium MSRs can deliver reliable electricity without adding carbon emissions. This positions thorium not just as a solution for today's grid but as a technology that prepares for the future dynamic grid.

Turning Seawater into Synthetic Fuel: Seafineries and the Hydrogen-Carbon Transition

Thorium reactors not only generate electricity and support saltwater purification but can also produce synthetic fuels from seawater. Countries with access to coastlines, thorium, and established nuclear infrastructure have the potential to become global suppliers of clean fuels. These systems can be combined with water purification processes, utilizing the same thorium reactor to generate energy for both freshwater production and synthetic fuel creation. Additionally, the reactor heat can be used for hydrogen production, maximizing the return on infrastructure investment.

These next-generation fuels, known as e-fuels, are a type of synthetic fuel made using carbon-free or low-carbon electricity, typically from renewable or nuclear sources to power the electrolysis of water and produce hydrogen. The "e" stands for electricity, highlighting their clean-energy origin. This hydrogen is then combined with carbon dioxide, either captured from the air or extracted from seawater to create liquid hydrocarbons like gasoline or other fuel. Unlike synthetic fuels made from fossil sources, e-fuels are carbon-neutral when burned, since the CO_2 released is the same CO_2 that was originally removed from the environment.

Imagine a seaside thorium MSR operating continuously, providing simultaneous high-temperature heat and stable electricity. Nearby, seawater undergoes electrolysis to generate hydrogen, while carbon dioxide is extracted from the ocean using acidification cells. The resulting hydrogen (H_2) and carbon dioxide (CO_2) are combined in a sea refinery, creating liquid hydrocarbon fuels like gasoline, jet fuel, and diesel.

TURNING SEAWATER INTO SYNTHETIC FUEL: SEAFINERIES AND THE HYDROGEN–CARBON TRANSITION

Thorium reactors not only generate electricity and support desalination but can also produce synthetic fuels from seawater.

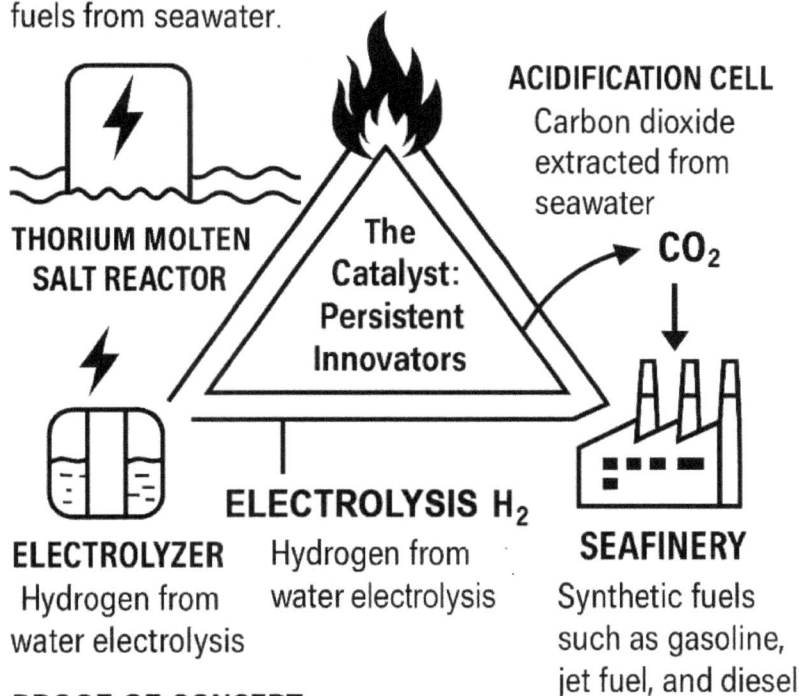

ACIDIFICATION CELL
Carbon dioxide extracted from seawater

THORIUM MOLTEN SALT REACTOR

The Catalyst: Persistent Innovators

CO_2

ELECTROLYSIS H_2

ELECTROLYZER
Hydrogen from water electrolysis

Hydrogen from water electrolysis

SEAFINERY
Synthetic fuels such as gasoline, jet fuel, and diesel

PROOF OF CONCEPT
In 2010, a U.S. Naval Research Laboratory study demonstrated the ability to produce synthetic jet fuel at sea from carbon dioxide and hydrogen.

Fuel, at its core, is simply hydrogen and carbon linked into chains of varying lengths. Short chains of one to four carbon atoms give gaseous fuels like methane (CH_4), propane (C_3H_8), and butane (C_4H_{10}). Medium-length chains such as hexane (C_6H_{14}) form liquid components of gasoline, while longer blends in the C_8–C_{16} range become jet fuel and diesel. Once we have a clean stream of hydrogen from electrolysis and carbon from seawater or captured air, making any of these fuels is a matter of choosing the right catalytic recipe and reactor temperature. In other words, the thorium-powered sea refinery can tailor its output, such as gas for cooking stoves, gasoline for cars, or kerosene for aircraft, simply by steering the chemistry toward the desired chain length.

Proof of Concept: The US Navy's Synthetic Fuel Breakthrough

In 2010, researchers from the US Naval Research Laboratory (NRL) published a study titled "The Feasibility and Current Estimated Capital Costs of Producing Jet Fuel at Sea Using Carbon Dioxide and Hydrogen." The study outlined a method to extract CO_2 from seawater, where CO_2 is more concentrated than in the air, and combine it with hydrogen to produce synthetic jet fuel. They demonstrated a modified Fischer–Tropsch process to create fuel performing similarly to JP-5, the US Navy's standard aviation fuel. The main challenge is the energy required for the process, necessitating a powerful, compact, and carbon-free energy source, which thorium reactors could provide. The research team concluded: "If the electricity and process heat required for carbon capture and hydrogen electrolysis can be provided by an onboard reactor, it would be possible to produce jet fuel autonomously and sustainably at sea."[6]

The Case for Thorium-Powered Seafineries

Building on this concept, Robert Hargraves, author of *New Nuclear is Hot In 2025*, expanded the idea of "seafineries".[7] These modular facilities make synthetic fuel and are placed next to advanced nuclear power plants. Hargraves explains that MSRs can provide both the electricity needed for electrolysis and the high-temperature heat that helps make the process more efficient. MSRs are compact and scalable, making them suitable for use at coastal desalination hubs, naval bases, or places that produce synthetic fuel.

A thorium-powered seafinery would offer a range of benefits:

Energy Security: Countries could produce aviation and transportation fuels domestically, without relying on imported oil.

Climate Mitigation: Synthetic fuels made from seawater are carbon-neutral when combusted, as the CO_2 released is the same CO_2 initially harvested from the ocean.

Defense Logistics: Naval fleets could refuel at sea or at coastal bases without tanker dependence, enhancing operational autonomy.

As global industries work to decarbonize aviation, shipping, and long-haul transport, synthetic fuels are set to become essential. The International Energy Agency predicts that e-fuels could account for up to 30% of aviation fuel by 2050. However, many existing synthetic fuels depend on intermittent renewable sources and face production challenges. Scaling up synthetic fuel production needs investment in infrastructure and clear regulatory frameworks, particularly for offshore reactors. However, the essential steps are already underway: in 2024, Indonesia signed a framework agreement with ThorCon to deploy a 500-MW thorium MSR aboard a moored vessel, explicitly designed to supply both electricity and desalinated water. This is a model that could be adapted to synthetic fuel production.

Thorium presents a viable solution for continuous, high-output synthetic fuel production, providing a clean and reliable energy source that does not compete with food crops or biomass. Furthermore, thorium-powered "seafineries" could effectively link the hydrogen economy with the existing global liquid-fuel infrastructure.

With synthetic fuel, clean water, and energy abundance fueled by thorium, the next chapter explores how these tools could do more than sustain economies; they could bring political stability and security to regions worldwide, perhaps even periods of peace. What if the solution to global instability wasn't just policy, but power?

1. DEPARTMENT OF ATOMIC ENERGY (INDIA), "KALPAKKAM HYBRID DESALINATION PROJECT," ANNUAL REPORT 2006–07 (MUMBAI: DAE, 2007), 112.

2. THORCON INTERNATIONAL, "THORCON AND PLN SIGN FRAMEWORK AGREEMENT FOR 500 MW TMSR-500," NEWS RELEASE, 6 FEBRUARY 2024.

3. EMIRATES NUCLEAR ENERGY CORPORATION, KHALIFA UNIVERSITY, AND FLIBE ENERGY, RUWAIS MSR–DESALINATION FEASIBILITY STUDY (ABU DHABI, 2023), EXECUTIVE SUMMARY.

4. NATIONAL RENEWABLE ENERGY LABORATORY. ADVANCED NUCLEAR AND GRID FLEXIBILITY: LOAD-FOLLOWING CAPABILITIES OF MOLTEN SALT REACTORS, NREL TECHNICAL REPORT, 2023.

5. US DEPARTMENT OF ENERGY. "MODELING LOAD-FOLLOWING PERFORMANCE OF ADVANCED REACTORS," OFFICE OF NUCLEAR ENERGY, GRID MODELING WORKSHOP, OCTOBER 2023.

6. HEATHER D. WILLAUER ET AL., "THE FEASIBILITY AND CURRENT ESTIMATED CAPITAL COSTS OF PRODUCING JET FUEL AT SEA USING CARBON DIOXIDE AND HYDROGEN," U.S. NAVAL RESEARCH LABORATORY REPORT NRL/6180-10-9046 (WASHINGTON, DC, 2010), 3.

7. ROBERT HARGRAVES, NEW NUCLEAR IS HOT IN 2025 (WHITE RIVER JUNCTION, VT: HARGRAVES, 2025), 61–65.

Chapter 7

Peace Through Power

Reimagining Global Energy Security

What if the key to peace wasn't a treaty but a power source?

For over a century, nations have fought over oil. They've jockeyed for gas pipelines and clashed over energy transit routes. Dependency has defined vulnerability, and energy scarcity has stoked conflict. But thorium could flip that equation. Instead of concentrating power in oil-rich regions, thorium could distribute energy sovereignty across the globe.

Thorium reactors, especially in modular forms, can be built domestically, sized to local needs, and moved near the point of use. That means fewer reasons to fight over pipelines, shipping lanes, or underground energy reserves. It also means that developing nations, too often left behind in the energy race, could leapfrog directly to clean, scalable power. When a country can produce its energy, it gains control over its destiny, making it less susceptible to the whims of global supply chains and price hikes. This autonomy increases stability and resilience, ensuring power remains available, especially in turbulent times.

Strength of Local Power: Thorium's Global Edge

Thorium reactors provide a local power advantage over traditional uranium reactors, requiring less infrastructure and water. They can be deployed on a smaller scale, enabling decentralized energy grids that produce stable, on-site power. This reduces transmission losses and enhances grid resilience. By using thorium's efficient and sustainable energy generation, countries can create environmentally friendly and economically viable energy systems, lowering carbon footprints and leading the global transition to clean energy.

Countries with Monazite Reserves and Thorium Potential

(*Darker means more Monazite Reserves*)

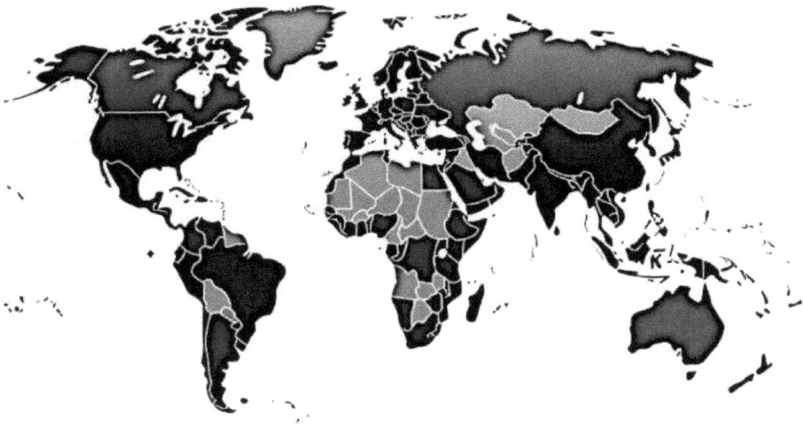

Local Power Plants Reduce Fossil Fuel Dependency

The benefits of energy independence go beyond just electricity generation. By reducing reliance on imported fuels, countries can improve their trade balances and redirect funds from energy imports to vital sectors like healthcare, education, and infrastructure. This shift can create jobs, exemplified by Iceland's geothermal systems[1] and France's nuclear infrastructure. Iceland uses underground heat energy to support its aluminum export industry and increase per-capita income. Meanwhile, France's nuclear program[2] has provided about 70% of its electricity since the 1990s, protecting consumers from volatile gas prices and keeping industrial energy costs low.

Thorium plants can readily combine with existing infrastructure, boosting economic growth. Molten-salt small modular reactors (SMRs) can be factory-built and connected to local grids with minimal upgrades. These reactors can function as self-contained microgrids in remote areas like Australian mining hubs, Alaskan military bases, or Caribbean resorts. They operate continuously, stabilizing power and reducing the energy interruptions and blackouts often seen with fossil fuel systems during extreme conditions.

Resilience also includes cybersecurity. Traditional power grids rely on a few large gas or coal plants and long pipelines, making them vulnerable. The 2021 Colonial Pipeline ransomware incident[3] highlighted how entire areas can lose power if one facility fails or a pipeline is compromised. A network of many smaller SMRs eliminates single points of failure: if one module pauses for maintenance, others continue to operate.

Yet in an increasingly digitized energy ecosystem, cybersecurity must be considered essential to resilience as physical durability or thermal stability. AI-managed reactors, remote monitoring systems, and grid-integrated SMRs introduce digital entry points that must be hardened against attack. A malicious breach could target sensor networks, disable remote shutdown protocols, or corrupt real-time data used to regulate salt chemistry in molten reactors.

The US Department of Energy and the Nuclear Regulatory Commission have begun issuing cybersecurity guidelines for digital nuclear systems, but many policies are still aimed at older, legacy plant systems.[4] As thorium SMRs and microreactors are deployed into critical civilian and defense infrastructure, including hospitals, desalination hubs, and industrial sites, they must be designed with strong cybersecurity. This means using encrypted command pathways, isolated networks, and AI fault detection to prevent manipulation or remote sabotage. In this way, cybersecurity cannot be an afterthought but must be integrated in the design as a foundational layer of trust in the nuclear future.

The US Department of Defense's Project Pele study highlights that a 5-MW molten-salt micro-reactor is capable of maintaining essential power during disruptions in regional transmission or fuel supply. This reliable and cost-effective energy source is vital for community well-being, ensuring that schools and hospitals operate smoothly, businesses can grow, and households have consistent power. A dependable power supply allows utilities to incorporate more variable renewable sources like wind and solar without compromising service quality. As countries implement thorium SMRs, they acquire a solution that reduces carbon emissions, mitigates fuel supply issues, and strengthens the grid against both physical and cyber threats, all while creating new job opportunities in advanced manufacturing and reactor operations.

Cooling Conflict: Reducing Tensions Through Energy Abundance

Competition over vital resources like oil, gas, and water often fuels instability and raises the risk of conflict. Struggles over access can quickly escalate, jeopardizing global peace. When nations depend heavily on a few key resources, they become vulnerable to market swings and supply disruptions, which can strain international relations. This reliance creates a fragile balance of power that shifts with energy prices and resource availability, leading to ongoing uncertainty and competition as countries fight for limited resources essential for modern life.

By offering an alternative to traditional fuels, thorium can help alleviate these tensions and ease resource competition. Thorium reactors could produce energy, purify water locally, and reduce the need for long-distance energy imports, minimizing the strategic vulnerabilities of reliance on foreign resources. This diversification strengthens national sovereignty and fosters a more cooperative global environment where energy becomes a tool for peace rather than conflict.

Case Studies in Clean Power: Regional Leaders and First Movers

Countries worldwide are turning to thorium to decrease dependence on imported energy and create more resilient, decentralized power systems. Countries across every inhabited continent have known thorium reserves, making it geopolitically accessible. This section will examine how regional leaders are moving to implement thorium-fueled power to ensure clean, scalable, and sustainable energy security.

First- Projects Mover

Country	Project Name / Reactor Type	Lead Organization(s)	Deployment Type	Thorium / MSR Relevance
United Arab Emirates (UAE)	Barakah Nuclear Power Plant (APR-1400)	Emirates Nuclear Energy Corp (ENEC), KEPCO	4 x 1400 MWe commercial reactors	Not MSR-based but establishes regional nuclear regulatory expertise and public trust critical for future MSRs.
Indonesia	TMSR-500 Floating MSR (ThorCon)	ThorCon, Ministry of Energy, PLN	500 MWe thorium MSR on barge platform	Directly applies thorium molten salt reactor (MSR) technology for coastal power and desalination.
China	TMSR-LF1 Prototype (Liquid Fuel MSR)	Chinese Academy of Sciences, SINAP	2 MWe experimental MSR (commissioned 2021)	First modern liquid-fueled thorium-compatible MSR; foundational for scaled-up future designs.
Canada	Moltex SSR-W and SMR Roadmap	Moltex Energy, NB Power, Canadian Nuclear Labs	Sited at Point Lepreau, New Brunswick	Salt-based reactor platform with future thorium pathways; uses recycled nuclear waste and stable salt cooling.

Southeast Asia is emerging as a key region of interest

Indonesia faces significant challenges in fuel delivery due to its 17,000 islands, which makes logistics complex and costly. To tackle this issue, the government is considering nuclear energy, particularly thorium-based SMRs. The Indonesian National Nuclear Energy Agency (BATAN) is interested in thorium molten salt reactors to fulfill future energy needs, especially on islands with unreliable fossil fuel transport. In 2023, Indonesia signed a memorandum of understanding with the US company ThorCon to develop a 500 MW thorium MSR that will be deployed as a floating barge, providing a novel solution for coastal energy supply.[5]

Thailand seeks to reduce reliance on natural gas, which currently makes up over 65% of the country's power generation. The Thailand Institute of Nuclear Technology (TINT) studies thorium fuel cycles with international partners. Thailand's Power Development Plan includes plans for advanced reactors. Researchers at Chulalongkorn University have also published studies on molten salt reactors using thorium. [6]

In Africa, interest in thorium is gaining momentum

South Africa holds some of the continent's largest thorium deposits, primarily from rare earth mining in the Western Cape and Northern Provinces. With its historical experience in nuclear energy from the Koeberg plant, the country has engaged in technical workshops for thorium reactor design and international collaborations on molten salt reactors. In 2022, the South African Nuclear Energy Corporation participated in discussions about small modular thorium reactors for rural electrification and industrial use.[7]

Namibia and Egypt are also exploring thorium as a part of their nuclear research. Namibia, a major uranium exporter, recognizes its thorium potential in heavy mineral sands and is incorporating advanced nuclear into its energy plans. These regional initiatives are part of a broader shift from fossil fuels towards cleaner, homegrown energy solutions that enhance sovereignty, grid resilience, and climate goals. Thorium's suitability for modular reactors and its ability to support desalination, hydrogen production, and synthetic fuels make it an attractive option for countries aiming to modernize their energy infrastructure with a clean energy future with flexibility and control.[8]

Peace Through Power: A New Vision for International Collaboration

Just as the Marshall Plan rebuilt Europe with steel and electricity, a "Peace Through Power" initiative could equip volatile regions with thorium-powered infrastructure: clean water, sustainable agriculture, and independent electricity.

This isn't idealism; it's a practical approach. When people have power, water, and food, insurgency becomes a harder sell. When local communities can own and operate their energy systems, trust builds.

International bodies like the IAEA, World Bank, or regional development banks could play a catalytic role in financing and coordinating these efforts, especially as thorium reactor technology matures and costs fall.

History shows us that shared energy efforts can bring old rivals together. After World War II, European countries formed the Coal and Steel Community[9], managing key resources jointly. That collaboration laid the groundwork for the European Union. In Southern Africa, cross-border energy projects have helped neighboring countries work together, proving that cooperation over energy isn't just possible; it's powerful.

Cross-border Collaboration

Thorium-fueled power and water purification plants could be the next step in cross-border collaboration. Thorium reactors can be built locally, sized according to the need, and don't carry the same weapons risks as uranium, making them ideal for international cooperation. Rather than competing for oil and gas, countries could work together to build and share thorium-powered systems, leading to cleaner energy and increased trust.

Imagine two nations with a complicated history collaborating on a thorium research center, creating a shared energy corridor that benefits both through a consistent power supply and economic growth. Scientists and policymakers are already discussing this type of collaboration.

Additionally, thorium can help lower global tensions related to fossil fuel dependence. Countries often clash over oil and gas access. Still, thorium presents a cleaner, more stable alternative for domestic production, aiding in water purification and reducing conflicts over water resources.

Thorium has brought nations together before. In his autobiography, *The First Nuclear Era*[10], Alvin Weinberg wrote:

"Our claims that nuclear power, if on a large-enough scale, was cheap enough to desalt the sea had reached the ear of Lewis Strauss [an investor and nuclear advocate]. The time was 1967, just after the Six-Day War, and the Middle East was much on our minds. The ultimate

problem of the Middle East is water - a point that is perhaps more fully recognized. . . Why not build several large, dual-purpose nuclear and desalination plants in Egypt, Israel, and Jordan - literally, to make the desert bloom - and thereby create a major new possibility for a settlement of the Israeli-Arab conflict?"

The 1967 Senate Resolution 155 called for a study[11] on nuclear-powered desalination in the Middle East, directed by the Atomic Energy Commission and conducted by Oak Ridge National Labs with engineers from Israel and Egypt. The study proposed "nuclear-powered agro-industrial complexes" to generate electricity and provide water for high-value crops near Alexandria, the Gaza Strip, and the western Negev. Still, those plans were never realized due to political challenges and the availability of cheap fossil fuels.

Investing in thorium projects and fostering international cooperation can turn energy into a source of stability and peace. This initiative goes beyond technology; it aims to create a future where nations collaborate for a cleaner, more prosperous world, starting with cross-border water purification hubs.

Global Collaboration Makes a Difference

International cooperation is essential for advancing thorium technology. The International Atomic Energy Agency (IAEA) facilitates research projects that unite scientists from China[12], India, Norway, and the United States to share insights and resources while aligning regulatory frameworks.[13]

Energy diplomacy also plays a role, enabling countries to establish agreements for sharing thorium research and expertise. For example, India[14] and Russia collaborate[15] on thorium fuel cycles, and China has welcomed international observers[16] and collaborators for its thorium MSR program.

Powering Progress: How Thorium Can Help Developing Nations

For many developing countries, energy isn't just about electricity; it's the foundation for progress. Without reliable power, growing an economy, running a hospital, or keeping schools open is nearly impossible. Frequent blackouts and unstable grids are everyday problems in many regions, holding back everything from basic services to business growth.

Thorium reactors present a practical solution. They can be sized to meet local needs, from small village units to larger city installations. They are also more cost-effective to operate than traditional nuclear plants and don't require extensive infrastructure, making them suitable for nations with budget constraints. Adopting this technology does come with challenges, primarily funding and technical expertise. However, solutions exist. International banks and development funds can offer the necessary capital, while training programs can equip local engineers to manage and maintain the reactors. Collaborations with countries experienced in nuclear energy can facilitate knowledge transfer.

With appropriate support, thorium could revolutionize energy access, providing clean, affordable power where needed most. This would create jobs, improved healthcare, stronger educational systems, and better opportunities. It also promotes equality, empowering less affluent communities to develop and compete globally.

Furthermore, as countries achieve energy independence, global stability, and connectivity increase, sharing thorium technology and expertise can strengthen international relations. Thus, thorium represents an energy solution and a pathway to a cleaner, fairer, and more collaborative world.

Powering and Watering Remote Communities with Thorium Technology

Remote communities often face energy challenges, relying on unreliable power sources and experiencing high diesel delivery costs. Thorium SMRs can provide stable, affordable energy and power reverse osmosis systems for clean water purification. This can improve agriculture in arid regions by facilitating irrigation and crop growth.

By ensuring reliable energy and water access, thorium technology can create jobs and stimulate economic growth in remote areas, enhancing food security and connecting them with urban markets. Innovative deployment of SMRs in these underserved regions presents a key opportunity for improving energy access and socio-economic conditions, leading to a brighter future.

Empowering Remote Communities with Microgrids

In Nunavut, Canada, Inuit communities rely on costly and polluting diesel shipments for heat and electricity, necessitating energy reform. The Canadian government has committed over CAD 1.6 billion to reduce diesel reliance in remote Indigenous communities through its Clean Energy for Rural and Remote Communities program. At the same time, Canadian Nuclear Laboratories (CNL) and Ontario Power Generation (OPG) are studying the possibility of using SMRs in off-grid northern regions, including Nunavut. Although there are no specific plans for thorium projects yet, SMRs, especially molten salt designs, are favorable for providing stable and low-maintenance energy in the Arctic's harsh environment. A thorium-powered microreactor could become a local energy source, lowering fuel costs and environmental risks while supporting heating, communication, and water purification.[17]

The Navajo Nation in the southwestern US looks to gain energy independence through clean nuclear technology. In 2023, the

154

Navajo Transitional Energy Company (NTEC) began exploring advanced nuclear solutions, including SMRs, to move beyond coal. Though not currently proposing thorium reactors, community leaders are interested in safe technologies focusing on land care and job creation. Turning abandoned uranium mine sites into thorium-powered research centers or microreactor locations would address past damage while providing clean, locally controlled energy.[18]

Working together with communities is key. Thorium reactor projects should be designed around local needs and in collaboration with tribal leaders.

Smart Policy for Smarter Power

So far, we've seen how thorium can decentralize power and foster global cooperation. But to make that vision real, policy must evolve. No matter the technology's promise, thorium energy won't scale without the right policies. Just as early wind and solar industries required targeted subsidies and permitting reform, thorium reactors need new frameworks that reflect their unique safety profile, design architecture, and deployment potential.

Clear and adaptive regulation is foundational. Most nuclear licensing regimes were built around conventional pressurized water reactors fueled by enriched uranium. These systems operate at high pressure and require massive containment structures. Thorium molten salt reactors, by contrast, run at atmospheric pressure, use liquid fuel, and incorporate passive safety mechanisms. To accommodate these differences, governments must modernize their nuclear codes, not by lowering standards, but by right-sizing them.

A growing number of jurisdictions are stepping up.

South Korea has invested over $1.2 billion since 2020 in its Next-Generation Nuclear Power Development Initiative, funding research on advanced reactors, including those compatible with thorium. The country's proactive regulatory structure and support for public-private consortia have made it a hub for nuclear innovation in Asia.[19]

In the United Kingdom, the Office for Nuclear Regulation (ONR) has overhauled its advanced reactor evaluation process, introducing a "technology-neutral" framework that allows molten salt reactor developers to propose alternative licensing pathways. This policy shift has made the UK a potential testbed for early thorium demonstration units.[20]

Canada is another standout. Through the Canadian Nuclear Safety Commission (CNSC), the country has created a Vendor Design Review (VDR) process that allows emerging reactor developers to receive feedback early in the development cycle, even before seeking full licensing. US-based thorium and molten salt developers like Terrestrial Energy have engaged with CNSC to help streamline their eventual deployment in North America.[21]

These efforts show that smart policy needs to reduce barriers and enable new developers. Policy experts say the best way the government can help is to modernize and streamline nuclear regulations so they are science-based, risk-informed, and supportive of innovation, while stepping back from subsidies and central planning. Empowering private industry and state-level oversight can create a competitive, market-driven nuclear energy revival. A liability insurance model that rewards safer reactor designs with lower premiums would further support new startups. Public-private partnerships, where researchers, startups, utilities, and regulators collaborate to de-risk first deployments, are also key to accelerating progress.

Licensing for the Next Generation

While thorium MSRs offer clear advantages in safety, waste reduction, and scalability, most nuclear regulatory frameworks, particularly in the United States, are still structured around 20th-century light-water reactor designs. These legacy regulations emphasize pressurized containment, solid fuel rod management, and coolant loop integrity, all of which are irrelevant or counterproductive when applied to molten salt systems.[22]

For thorium MSRs to move from pilot projects to commercial deployment, regulatory reform is essential. The US Nuclear Regulatory Commission (NRC) has taken steps toward performance-based, technology-inclusive licensing through its Advanced Reactor Content of Application Project (ARCAP), but progress is slow.[23]

Key reforms needed include:

1. creating licensing pathways for reactors that operate at atmospheric pressure and use liquid fuels,
2. updating safety case evaluation methods to accommodate passive shutdown mechanisms and salt chemistry, and
3. allowing staged or modular certification approaches that fit SMR deployment models.[24]

Internationally, regulators in Canada and the UK have shown more flexibility, with pre-licensing review processes now open to non-traditional designs. Still, for thorium MSRs to scale globally, regulatory cooperation and harmonization, particularly in areas such as fuel handling, non-proliferation compliance, and site risk assessments, must become priorities. Without these reforms, the pace of innovation will continue to lag behind the promise of the technology.

Finally, to scale globally, thorium energy needs international harmonization. Shared safety standards, materials databases, and operator training protocols can foster trust across borders. Initiatives like the Generation IV International Forum[25] and the International Atomic Energy Agency's SMR Regulators Forum[26] are already laying the groundwork for cross-border technical validation and export compatibility.

In short, the technology is ready, but the rules must catch up. With wisely adaptive policy frameworks, thorium can move from lab bench to launchpad, offering countries a clean, reliable, and secure alternative to fossil fuels. And in the process, policy itself becomes a driver of innovation.

How You Can Help: Grassroots to Global Action

When we think about shaping the future of energy, it's easy to assume power lies solely with government officials, energy executives, or big corporations. But throughout history, some of the most important shifts have been sparked by everyday people, citizens who cared enough to speak up. From taxpayer rights to environmental protection, grassroots movements have changed the course of history. Today, many people are raising their voices to support thorium energy as a clean, safe, and reliable solution to our global energy and water challenges.

From Local Conversations to Global Change

In communities across the world, local groups start conversations, organize events, and share information about thorium. They're educating neighbors, challenging outdated assumptions about nuclear power, and creating momentum for change. These grassroots efforts build public understanding and support, which are essential if thorium is to be taken seriously by those in power. As awareness spreads, more people realize that thorium offers a way to produce clean energy and a chance to generate fresh water and reduce our reliance on fossil fuels.

Turning Support Into Policy

But raising awareness is just the beginning. To bring thorium into the mainstream, public support must reach policymakers who write the rules, fund the projects, and shape our energy future. That means citizens need to engage directly with their representatives, attend public forums, submit feedback on energy plans, and make their voices heard. When decision-makers see genuine enthusiasm and understanding around thorium as integral to a clean energy policy, they're more likely to support research funding, regulatory updates, and long-term investment in this promising technology. Public advocacy becomes not just about pushing for change but about working alongside leaders to build policies that reflect the will and needs of the people.

Meet the Organizations Making it Happen

Several advocacy groups are already leading the way:

Thorium Energy Alliance (TEA) brings together scientists, engineers, and citizens to promote thorium technology. They host conferences, provide educational resources, and advocate for policy changes in Washington and beyond.[27]

Nuclear Innovation Alliance (NIA) works with energy experts and government agencies to support new reactor designs, including thorium-based systems.[28]

EnergyFromThorium.com is a grassroots hub with resources, news, and discussion forums where people can learn, connect, and organize.[29]

These groups offer ways to get involved, whether you want to volunteer, donate, share content, or attend a local event.

Using the Power of Digital Advocacy

Today, public advocacy often happens online. Social media platforms, blogs, videos, and forums give people a powerful way to share knowledge, spread ideas, and connect with like-minded individuals worldwide. A single post or article can reach thousands, sometimes millions, of people in hours. This digital momentum turns isolated supporters into a networked movement, creating the kind of public visibility that makes them hard to ignore.

Global Lessons from Local Action

Public support can significantly influence national energy strategies. For instance, citizen advocacy fueled Germany's shift[30] toward renewable energy, leading to major wind and solar power investments. Similarly, strong public backing in France[31] has maintained a mostly carbon-free power grid, and that advocacy can now drive interest in safer technologies like thorium.

These examples highlight the power of an informed and organized public in shaping energy policy. Today, there's a chance to advocate for thorium. You don't need to be a scientist or policymaker to make an impact. Whether writing articles, attending local meetings, posting on social media, or discussing with friends and family, your voice contributes to the growing demand for cleaner, safer energy. Real change occurs when individuals unite around a common goal and take action.

Turning Policy into Power: US Reforms Are on the Way

In the United States, public perception of advanced nuclear power is shifting, and long-overdue policy reforms are finally beginning to take hold.

States Are Opening up to Advanced Nuclear Power

States are taking legislative action due to environmental concerns, energy reliability issues, aging fossil infrastructure, and a desire to attract high-tech jobs and clean investment. State legislatures recognize that next-generation nuclear, including thorium-fueled designs, offers a rare combination of resilience, scalability, and economic potential.

Here are examples of nuclear energy reforms across the states:

Colorado: The US state of Colorado made history in March 2025 by passing House Bill 1040, which amended the state's definition of "clean energy resources" to explicitly include nuclear power and advanced reactor technologies. Notably, it is one of the first state statutes to name thorium as an eligible fuel type. This legal change enables thorium reactor projects to qualify for clean energy incentives, grants, and state procurement targets under Colorado's aggressive decarbonization strategy. For innovators and investors, the bill sends a clear message: Colorado intends to be a first mover in nuclear innovation, treating thorium as a peer to solar, wind, and hydrogen.[32]

Alabama: In April 2025, Alabama followed suit with bipartisan passage of the Advanced Nuclear Energy Facilitation Act. The law streamlines siting and permitting for next-generation reactors, including molten salt and thorium-based designs, by creating a state-level review panel that works alongside the Nuclear Regulatory Commission (NRC). Alabama's law also provides a property tax exemption for first-of-a-kind demonstration reactors and allocates state development bonds for microreactor and small modular reactor projects. The state's aerospace and materials engineering legacy has made it an attractive destination for thorium MSR developers, including those looking to manufacture reactor components at scale.[33]

Wyoming and Montana: Both states are encouraging the repurposing of coal plant sites[34] for small modular reactors. Legislation passed in 2024 and early 2025 facilitates quick environmental reviews and funding for coal communities transitioning to clean nuclear energy. While not specific to thorium, these states have created flexible regulatory space that thorium developers can leverage, particularly for modular molten salt designs.[35]

Idaho: Home to the Idaho National Laboratory (INL), the state has proposed tax credits and grant matching for private sector reactor trials at INL. Several thorium startups are discussing pilot projects at the lab.[36]

Tennessee: Home to Oak Ridge National Laboratory, birthplace of the thorium molten salt reactor[37], Tennessee lawmakers introduced legislation in 2025 to establish an Advanced Nuclear Commercialization Zone, offering fast-track permitting for nuclear startups and access to retired industrial sites for testing, and connecting universities to clean energy commercialization funding.[38]

Emerging developments: Indiana[39] and Ohio[40] are exploring how to include advanced reactors in their grid plans, while Texas[41] is considering including nuclear energy in its clean hydrogen and carbon capture frameworks. Even California[42], historically resistant to nuclear power, is under pressure to reconsider its moratorium due to carbon goals and grid reliability concerns.

In Summary:

The US is moving toward a new era of nuclear power featuring molten salt reactors, thorium fuels, and decentralized systems. States are implementing policies that promote innovation, quicken demonstrations, and attract private investment. These reforms aim to enhance job growth, economic development, and leadership in the global clean energy landscape.

US Nuclear Power Insurance and the Regulatory Straitjacket

Despite its potential for clean energy, the US nuclear industry faces significant insurance challenges that hinder innovation and deter private investment. Unlike other industries, nuclear power operators cannot obtain commercial accident liability insurance. Instead, they rely on the outdated Price-Anderson Act, established in 1957.

This act provides a federal backstop for liability coverage, formed due to private insurers' reluctance to cover the vast financial risks of nuclear accidents. It creates a two-tier system: the first tier involves coverage through private insurers affiliated with American Nuclear Insurers (ANI), and the second tier is a federal indemnity pool supported by mandatory contributions from nuclear plant operators up to a defined cap.

While intended to protect the public and promote nuclear development, the Price-Anderson Act has stifled advancements in reactor technologies like thorium-fueled MSRs. The insurance system is closely tied to older light-water reactor risks, disregarding the inherent safety of newer designs, leaving innovators like Flibe Energy in regulatory uncertainty, unable to demonstrate or insure their designs on a commercially viable timeline.

The insurance system is still based on the risks of older nuclear reactors, which is why Kirk Sorensen, the founder of Flibe Energy, has called for change. At several Thorium Energy Alliance conferences, he has drawn comparisons to the maritime shipping industry, where ship owners can choose between several insurers, such as Lloyd's of London, based on competitive rates and assessed risk. "Why can't we do that with reactors?" Sorensen asked at the TEAC7 conference. "If your design is safer, your premiums should be lower. That would finally align market incentives with safety innovation."[43]

Jack Spencer, a senior research fellow for nuclear energy policy and author of *Nuclear Revolution*, highlights that Price-Anderson is now outdated and discourages private risk evaluation and innovation by centralizing all liability through the government. "We need a free market in nuclear insurance," Spencer said at a panel in 2023. "The absence of competition means there is no reward for making a reactor genuinely safer."[44]

It's said that "if you can't insure it, you can't invest in it", but some progress is underway. In 2025, as part of the broader nuclear modernization effort, the Department of Energy and the Nuclear Regulatory Commission (NRC) have begun consulting with private insurers and reinsurance groups on how to evolve the liability framework. One proposal would allow advanced reactor developers to work with international insurers for primary liability, with federal reinsurance as a backup.

However, reform is politically challenging. Critics express concern about transferring risk to consumers, while supporters warn that a failure to modernize could result in losing nuclear leadership. In response to ongoing barriers like outdated insurance models and regulatory uncertainty, the federal government has implemented executive actions to facilitate the deployment of advanced reactors, including thorium-fueled ones.

US Executive Orders and the Thorium Opportunity

While US state legislatures are clearing the path for advanced nuclear energy from the bottom up, federal action is now pushing from the top down. A series of executive orders marked a pivotal shift in US nuclear policy, reframing nuclear energy not as a regulatory burden but as a national security asset and a cornerstone of energy resilience. These directives target long-standing roadblocks: outdated licensing processes, fuel supply bottlenecks, and insurance models that favor legacy designs over safer, modular innovations. Though none mention thorium by name, their impact could be transformative for molten salt reactor developers, creating new pathways to deploy reactors that were once sidelined by bureaucracy rather than science.

In May 2025, a series of presidential executive orders[45] collectively reoriented the trajectory of American nuclear policy, starting with a sweeping reform of the Nuclear Regulatory Commission (NRC), long criticized for its sluggish pace, unclear requirements, and a one-size-fits-all approach to diverse reactor technologies.

Executive Order 1: NRC Licensing Reform

This order establishes fixed timelines for NRC licensing: 18 months for new reactor applications and 12 months for renewals. It mandates adopting science-based radiation limits and streamlining NEPA compliance while creating a dedicated Advanced Reactor Office.

Impact: Science-based radiation standards are essential for the future of clean nuclear energy. For decades, outdated safety limits, often based on overly cautious Cold War-era models, have inflated costs, blocked promising sites, and distorted public perception about how safe today's advanced reactors really are. Just as importantly, we need clear and fixed licensing timelines. Right now, next-generation reactors often get stuck in years of regulatory limbo because the rules they're judged by were designed for older systems, specifically, pressurized water reactors. That uncertainty drives away investors, stalls innovation, and makes it difficult for new reactor developers to plan and build with confidence.

This has been a major barrier for thorium molten salt reactor developers. Their technology works fundamentally differently from legacy reactors - it runs at low pressure, uses liquid fuel, and includes passive safety features that don't require external power to shut down. But until recently, these reactors were still being evaluated using standards built for solid-fuel, high-pressure systems.

To fix that, the U.S. Nuclear Regulatory Commission has announced the creation of a dedicated Advanced Reactor Licensing Office. This new office is tasked with reviewing what are known as Generation IV designs, reactors that go beyond traditional nuclear in both form and function. These include molten salt reactors, gas-cooled reactors, fast-spectrum systems, and other designs that offer significant improvements in efficiency, safety, and waste reduction. By giving them their own regulatory lane, the NRC can evaluate these designs based on their actual technology, not by trying to retrofit old rules onto new innovations. This shift has the potential to dramatically reduce approval times and help bring thorium-powered solutions to market more quickly and affordably.

Executive Order 2: Nuclear for National Security

This directive tasks the Department of Defense with deploying a licensed small modular reactor at a military installation by 2028. It also instructs the Department of Energy to prioritize SMRs for powering AI data centers designated as critical defense infrastructure.

Impact: This creates a clear need for small modular reactors (SMRs) that can be safely used, a sweet spot for thorium-fueled liquid salt reactors, which run at normal air pressure and use heat very efficiently. It may lead to early federal procurement of thorium MSRs for secure military base and data center operations.

Executive Order 3: Fuel Supply Resilience and Recycling

This order invokes the Defense Production Act to release 20 metric tons of high-assay low-enriched uranium (HALEU). It encourages public-private partnerships for recycling spent fuel, including plutonium-containing materials.

Impact: High-assay low-enriched uranium, enriched between 5 and 20 percent U-235, is crucial for many advanced reactor designs, including those using uranium as a starter fuel for thorium breeding. It provides a higher neutron economy, so reactors can have smaller cores, use fuel more efficiently, and adjust their power output more flexibly to match the need.

The US doesn't yet have a robust supply chain for HALEU-grade uranium, so the planned release of 20 tons from federal stockpiles in 2025 will be a crucial boost for early reactor rollouts. It also provides breathing room for developers to bring online commercial HALEU enrichment facilities, which are expected to start production by 2027. The release signals a shift toward ensuring fuel supply diversity and infrastructure for advanced nuclear.[46]

Notably, the order calls for feasibility studies on converting surplus weapons-grade plutonium into usable reactor fuel, a promising avenue for molten salt reactors designed to handle a wide range of nuclear materials. Traditionally viewed as a security risk, surplus plutonium can be utilized in fast-spectrum MSRs. Proposed regulatory updates may allow the controlled use of surplus plutonium, aligning with nonproliferation goals. This would not mean universal plutonium recycling, but rather licensing pilot reactors that consume it safely within sealed, monitored fuel cycles. Once plutonium is used as fuel, it is no longer a threat. Turning a long-term storage liability into a near-term energy asset could shift the narrative from nuclear risk to nuclear responsibility.

In Summary:

The 2025 executive orders and associated DOE reforms have opened the door for a new class of nuclear technologies, especially thorium MSRs. By attacking the delays in licensing, the bottlenecks in fuel supply, and the fragility of insurance coverage, these reforms signal that the era of regulatory hostility may finally give way to a period of rational nuclear enablement.

The Way Forward: A New Architecture for Peace

This chapter outlined how thorium technology, effective policies, and global collaboration can enhance energy supply and promote peace. Decentralized thorium reactors allow countries to achieve energy independence, avoiding the pitfalls of fossil fuel reliance. This leads to local power production, economic stabilization, and investments in health and education. Case studies from Southeast Asia, Africa, the Arctic, and the American Southwest demonstrate that this approach to energy is gaining momentum as a growing movement.

For investors with portfolios still heavily tied to fossil fuels, thorium presents not a threat but a hedge as a prudent way to diversify energy assets as the global economy shifts toward decarbonization. The logic is straightforward: while oil and gas will remain relevant for years, their long-term viability is increasingly limited by climate mandates, accelerating electrification, and evolving public sentiment. Savvy petro-states are already adapting. The United Arab Emirates has invested billions in nuclear infrastructure, including a feasibility study with Flibe Energy on thorium-powered desalination.[47] Saudi Arabia's K.A.CARE program has investigated thorium as part of its broader strategy to transition from oil revenue to sustainable energy leadership.[48] These moves aren't ideological; they're financial. For energy-sector investors, backing thorium now, through advanced reactor start-ups, industrial applications, or utility partnerships, offers a foothold in a market with potentially decades of compounded returns. The same wealth that once came from hydrocarbons may, in time, flow from thorium-fueled heat and power.

On the international stage, thorium's compatibility with nonproliferation and grid stability fosters cooperation rather than competition, creating potential bridges between nations, as seen from historical examples to current partnerships.

However, envisioning this future requires action. The chapter also highlighted the challenges of outdated regulatory frameworks and how recent US executive orders are improving the market. Changes in NRC licensing timelines increased HALEU availability, and initiatives for safe plutonium management are setting the stage for a new nuclear landscape with new opportunities for thorium-fueled startups.

NUCLEAR GOVERNANCE PATHWAYS FOR MODERN MSRS

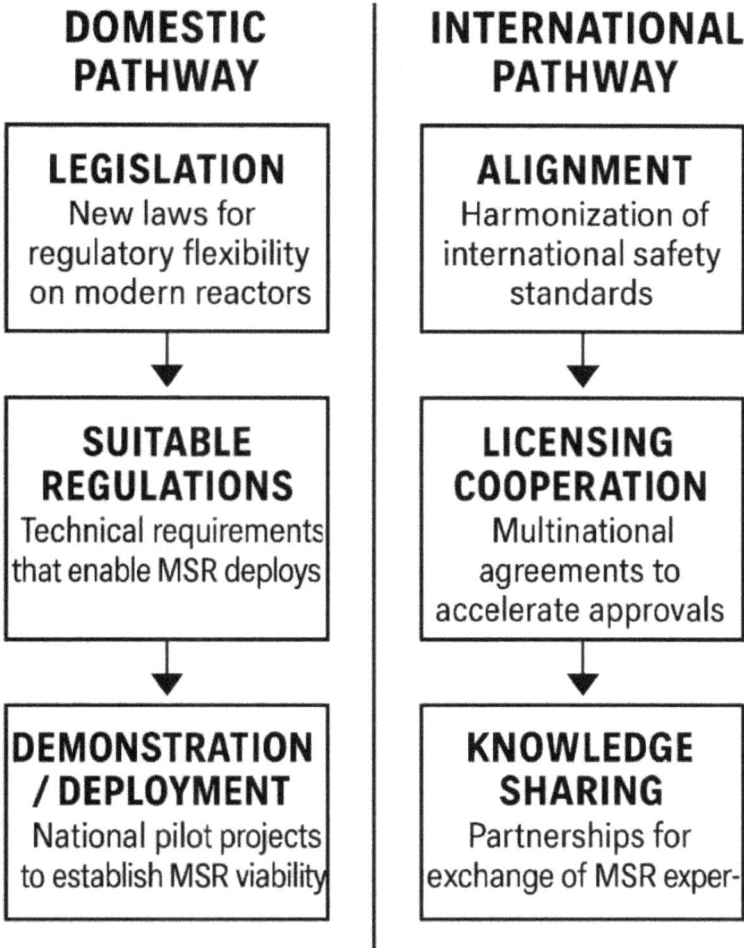

DOMESTIC PATHWAY

LEGISLATION
New laws for regulatory flexibility on modern reactors

↓

SUITABLE REGULATIONS
Technical requirements that enable MSR deploys

↓

DEMONSTRATION / DEPLOYMENT
National pilot projects to establish MSR viability

INTERNATIONAL PATHWAY

ALIGNMENT
Harmonization of international safety standards

↓

LICENSING COOPERATION
Multinational agreements to accelerate approvals

↓

KNOWLEDGE SHARING
Partnerships for exchange of MSR exper-

Energy can drive conflict or cultivate peace. When wisely deployed, it becomes a foundation for cooperation, resilience, and prosperity.

Beyond power grids and diplomacy, thorium's real promise may lie in its ability to sustain life itself by delivering freshwater, enabling year-round irrigation, and powering decentralized food systems. The next chapter explores how thorium energy could help feed a growing planet.

1. Government of iceland, national energy authority, "geothermal energy in iceland," 2023, https://nea.is/geothermal.

2. World nuclear association, "nuclear power in france," march 2024, https://www.world-nuclear.org/information-library/country-profiles/countries-a-f/france.aspx.

3. Cybersecurity and infrastructure security agency (cisa), "colonial pipeline ransomware attack," 2021 incident summary.

4. Us nuclear regulatory commission. Cybersecurity regulatory guide for digital instrumentation and control systems, rev. 1, january 2024. Https://www.nrc.gov

5. Thorcon international, "thorcon and pln sign framework agreement for 500 mw tmsr-500," news release, february 6, 2024

6. Thailand institute of nuclear technology (tint), "thorium research collaboration," policy brief, 2024.

7. South african nuclear energy corporation (necsa), "advanced reactor workshop summary," 2022.

8. World nuclear news, "namibia, egypt explore thorium research," july 2023.

9. European union, "history of the european coal and steel community," official records archive, 2023.

10. Alvin weinberg, the first nuclear era: the life and times of a technological fixer (new york: aip press, 1994), 202.

11. Us senate, "senate resolution 155: nuclear desalination study," congressional record, 1967.

12. World nuclear news, "china opens msr program to international collaboration," november 2024.

13. Iaea, "international research collaborations on advanced nuclear," 2024 progress summary.

14. Department of atomic energy (india), "kalpakkam hybrid desalination project," annual report 2006–07 (mumbai: dae, 2007), 112.

15. World nuclear news, "india-russia thorium fuel partnership advances," january 2025.

16. World nuclear news, "china opens msr program to international collaboration," november 2024.

17. Natural resources canada, "energy support for remote communities," 2024 program brief.

18. Navajo transitional energy company (ntec), "nuclear transition plan," 2023 announcement.

19. Korean ministry of energy, "next-generation nuclear power development initiative annual report," 2024.

20. Uk office for nuclear regulation, "technology-neutral advanced reactor licensing," regulatory update, 2023.

21. Canadian nuclear safety commission, "vendor design review program overview," 2024.

22. Per peterson, jacopo buongiorno, and robert hargraves. "why licensing for molten salt reactors must change." Nuclear engineering review, fall 2023.

23. Us nuclear regulatory commission. Advanced reactor content of application project (arcap), office of new reactors, 2024. Https://www.nrc.gov/reactors/new-reactors/advanced/arcap.html

24. Nuclear innovation alliance. "enabling nuclear innovation: strategies for modernizing nuclear licensing." April 2023. Https://www.nuclearinnovationalliance.org

25. Generation iv international forum, "global regulatory harmonization initiatives," 2023.

26. Iaea, "smr regulators forum annual report," 2024.

27. Thorium energy alliance, "about us," https://thoriumenergyalliance.com.

28. Nuclear innovation alliance, "mission and impact," 2024 overview.

29. Energyfromthorium.com, "learning resources and community discussions," 2025.

30. German federal ministry for economic affairs and climate action, "energiewende: policy and progress," 2024.

31. World nuclear association, "france: public support for nuclear energy," 2023.

32. Colorado general assembly, "house bill 1040: clean energy reform," 2025.

33. Alabama legislature, "advanced nuclear energy facilitation act," act no. 2025-112.

34. Wyoming department of energy, "coal-to-nuclear transition support programs," 2024.

35. Montana legislature, "sb2024-14: modular nuclear support act," 2024.

36. Idaho state government, "inl partnerships and incentives," economic development summary, 2025.

37. Oak ridge national laboratory, molten salt reactor experiment final report (1972).

38. Tennessee department of economic and community development, "advanced nuclear commercialization zone proposal," 2025.

39. Indiana legislative services agency, "nuclear grid study committee report," 2025.

40. Ohio house of representatives, "advanced reactors for resilience bill," 2025.

41. Texas energy commission, "public comment request: clean hydrogen framework," 2025.

42. California public utilities commission, "nuclear moratorium review brief20.

43. Kirk sorensen, remarks at thorium energy alliance conference 7 (teac7), 2016.

44. Jack spencer, nuclear revolution: powering the next generation (washington, dc: heritage foundation, 2020).

45. White house, executive orders 14089, 14090, and 14091, may 23, 2025, https://www.whitehouse.gov.

46. Us department of energy, "haleu availability and commercialization strategy report," march 2025.

47. Emirates nuclear energy corporation. "uae—flibe energy thorium desalination feasibility study." Project summary, 2024. Https://www.enec.gov.ae/en/media/news/uae-flibe-thorium-desalination-feasibility

48. King abdullah city for atomic and renewable energy (k.a.care). "strategic energy transition report." Riyadh, 2023. Https://www.kacare.gov.sa/en/strategy/energy-transition-report

Chapter 8

Thriving Agriculture

A Showcase for Thorium's Environmental Promise

Imagine a future where the air is fresher, the oceans are cleaner, and farmland worldwide flourishes without polluting the planet to get there. Thorium-powered energy systems have the potential to deliver exactly that, offering a path where clean energy, clean water, and sustainable food production reinforce one another.

The United Nations projects that the global population will exceed 9.7 billion by 2050[1], with most of this growth occurring in regions already facing food and water stress. To meet future demand, the Food and Agriculture Organization (FAO) estimates that global food production will need to increase by at least 60%[2] compared to 2005 levels. Achieving this without using more farmland or harming the environment will require major advances in crop yields, smarter use of resources, and resilient production systems, especially in regions facing water scarcity and energy constraints.

Climate change, water scarcity, and food insecurity are pressing challenges for communities worldwide. The future of agriculture relies on effective energy solutions. We need a future where power, water, and food security management are integrated. Thorium reactors can help enable a sustainable system supporting clean energy, clean water, clean fuels, fertilizers, and food production.

Farming the Future: Thorium and Clean Agriculture

Agriculture is one of the most energy-intensive industries on Earth. It requires massive amounts of electricity for irrigation, fertilizer production, food processing, and cold storage. It also depends heavily on transportation fuels for tractors, harvesters, and trucks. Today, much of this energy demand is met by fossil fuels, directly and significantly contributing to carbon emissions, soil degradation, and water shortages.

Thorium energy provides a clean and scalable solution through small modular molten salt reactors (SMRs). These reactors can power remote agricultural hubs, provide reliable water desalination and irrigation electricity, support sustainable fertilizer production, and co-produce synthetic fuels for farm machinery while significantly lowering environmental impact.

How MSRs Support Greenhouse Heat, LED Lighting, and Water Purification

GREENHOUSE HEAT	LED LIGHTING	WATER PURIFICATION
MSRs provide large amounts of heat for maintaining optimal growing conditions	Excess MSR power can drive efficient LED growth lighting	MSRs enable water purification methods like reverse osmosis

Agriculture and Water: Powering Food Security

Modern farming relies heavily on energy for tasks like pumping water for crops, powering indoor farms, and keeping food cold in storage. Farms using thorium reactors can generate clean, abundant energy, decrease pollution and improve precision agriculture. They can supply reliable electricity for advanced irrigation systems, ensuring optimal water use while minimizing waste. Additionally, these reactors can support food processing and storage, preserving produce quality from farm to table while improving efficiency, lowering costs, and boosting productivity in the agricultural sector.

Stable Water Supply Through Thorium-Powered Desalination

Agriculture's lifeblood is water. Yet, irrigation-dependent farming often competes with cities and industries for scarce freshwater supplies. Thorium reactors, operating continuously, can power reverse osmosis (RO) desalination plants to provide abundant fresh-

176

water even in dry regions. Molten salt reactors generate both electricity and high-quality heat, which is ideal for driving multi-stage water purification systems. Multi-effect distillation (MED) systems use less energy than simple boiling methods, making them particularly suitable for agriculture.[3]

In a MED system, salty or slightly salty water is turned into steam and then cooled back into freshwater across several connected low-pressure chambers. Each chamber reuses the heat from the previous one, which significantly lowers the energy needed to produce each liter of fresh water. Additionally, thorium reactors can shift heat between making electricity and producing water, responding to power and irrigation needs as they change in real time.

Thorium reactors provide reliable, continuous energy, unlike solar and wind systems that depend on weather. This reliability supports year-round irrigation, helping prevent soil damage caused by too little or too much water. It allows farmers to plan crop rotations and planting schedules confidently. In a 2015 article, the International Atomic Energy Agency emphasized nuclear desalination, especially with reverse osmosis, as a sustainable agricultural solution in drought-prone areas.

Food Resilience and Year-Round Growing

Combining thorium small modular reactors with advanced agricultural practices can enhance food resilience. A 20- to 50-MW thorium SMR placed next to greenhouses can provide continuous electricity for LED lights and water pumps, along with leftover heat to keep crops warm around 25 °C even in cold weather. This stable energy supply avoids the crop losses that can happen with solar or diesel power.

Reliable steady power makes it possible to use high-density hydroponics, an advanced growing method of compact water-based farming, adding extra CO_2 for plant growth, and fully automated temperature control, helping plants grow faster and healthier. In closed systems, CO_2 taken from reactor exhaust or nearby synthetic fuel-making facilities can be sent into indoor farms, helping plants grow up to 30% faster, according to research in controlled farming environments.[4]

In areas such as the Arctic or salty coastal flats, where soil is scarce, thorium reactors paired with RO water purification can turn slightly salty or ocean water into fresh water for irrigation, bringing life back to previously barren land. While this model holds promise for cold, remote regions like the Arctic, it is equally relevant in hot, dry zones, where water scarcity, rather than temperature, limits agricultural productivity. Successful test projects in the Middle East and India show that one liquid-salt reactor can provide both heat for indoor farming and power for a large water-purifying plant at the same time.[5]

Notably, nuclear power stabilizes the food supply against climate disruptions. Even during droughts or wildfires, the reactor's consistent output maintains photosynthetic light levels and keeps nutrient solution circulating, preventing crop losses that can devastate remote communities. Ultimately, thorium energy supports year-round farming in challenging environments.

Clean Fertilizer Production

Making fertilizer is one of the most carbon-intensive industrial processes in the world. The traditional method of producing ammonia, known as the Haber-Bosch process, relies heavily on natural gas. This gas isn't just used to power the process; it's also broken apart to create hydrogen, releasing large amounts of carbon dioxide in the process. For every ton of hydrogen produced, this method emits roughly 9–12 tons of CO_2, making it one of the most carbon-polluting industrial processes on Earth. In fact, fertilizer production accounts for more than 1% of global carbon emissions.

Thorium molten salt reactors offer a cleaner path. These high-temperature reactors, which operate around 700 to 800 degrees

Celsius, can provide both electricity and heat to produce hydrogen in a cleaner way, by splitting water instead of burning fossil fuels. This technique, called high-temperature steam electrolysis, uses advanced solid oxide electrolysis cells to separate water into hydrogen and oxygen more efficiently.[6] That hydrogen is then combined with nitrogen from the air to make ammonia—the same end product, but with a much lower carbon footprint. By replacing natural gas with thorium-powered hydrogen, fertilizer plants could slash emissions, cut energy use, and help feed the world more sustainably. The U.S. Department of Energy's Idaho National Laboratory has endorsed this approach, and in 2021, the DOE's Advanced Manufacturing Office named high-temperature reactors a key technology for low-carbon fertilizer production.[7]

Crop Yield Improvements from Stable Water and Fertilizer Access

Crop Type	Baseline Yield (tons/ha)	With Stable Irrigation	With Irrigation + Fertilizer	Typical Yield Increase
Maize (corn)	2.5–3.5	5.0–6.0	6.5–8.0	2.0–3.0×
Wheat	2.0–3.0	4.0–5.0	5.5–6.5	2.0–3.0×
Rice (paddy)	4.0–5.0	6.0–7.0	7.5–8.5	1.5–2.0×
Tomatoes	20–30	40–45	50–60	2.0–3.0×
Potatoes	10–15	18–22	25–30	2.0–2.5×
Soybeans	1.5–2.5	2.8–3.5	3.5–4.5	1.5–2.0×
Millet / Sorghum	1.0–1.5	2.0–2.5	2.8–3.5	2.0–3.0×

Local Food Processing and Storage

In Alaska, small modular thorium reactors paired with hydroponic greenhouses could effectively tackle food insecurity in areas like the Yukon-Kuskokwim Delta, where food costs exceed 200% of the US average[8]. These reactors would provide reliable heat and electricity year-round, boosting local food production, creating jobs, and alleviating energy challenges. Rural villages face high diesel costs and limited fuel access, prompting attention from state officials and national agencies.

The US Department of Energy (DOE) and the Alaska Center for Energy and Power (ACEP) are investigating microgrids and advanced reactors suited for the region. In 2022, the DOE identified Alaska as a promising site for testing microreactors[9], supporting planning for small reactors in local communities. Microreactors will enable greenhouse and hydroponic food solutions to supplement the local food supply.

Small reactors can power indoor farms and water-based growing systems to help add to the local food supply. This hot energy helps maintain stable soil and air temperatures, allowing for year-round crop cultivation and warming the water for fish farms. Aquaponic fish and plant growing systems can thrive with this warm water, enabling fish farming and vegetable growth under consistent conditions. Temperatures must stay between 20°C and 28°C to continuously support plant growth and fish metabolism. Reactor-supplied waste heat can maintain these ideal conditions even in harsh climates.[10]

The microreactor solutions that will improve food production in Alaska can be applied across rural communities, promoting a circular agro-energy economy where waste heat is repurposed to reduce food spoilage and resources are optimized to support sustainable food production.

Co-Production of Synthetic Fuels for Farm Equipment

Agriculture requires liquid fuels for tractors, combines, and logistics, not just electricity. Thorium reactors, located near the sea or inland brine sources, could power synthetic fuel plants by combining hydrogen produced using electricity from electrolysis with carbon dioxide extracted from seawater or the air.

This builds on the earlier idea of a sea refinery, Robert Hargrave's approach to make clean fuel in smaller, local setups. It uses a method called Fischer Tropsch synthesis, which works at temperatures between 200°C and 350°C.[11] The secondary loop of thorium molten salt reactors (MSR) can directly supply this heat, making it more efficient and simpler by removing the need for complicated heat equipment.

For farmers, the co-production of electricity plus man-made fuel would mean access to a local, clean, and reliable fuel supply, reducing reliance on volatile oil markets and carbon-heavy refineries. Environmentally, it allows for the recycling of carbon waste into cleaner fuel sources.

Agriculture Without Depletion

Thorium power can lead to an agricultural revolution focused on renewing natural resources. Agriculture can thrive without exhausting resources by using plentiful clean water, low-carbon fertilizers, local energy sources, and clean fuels.

Stable irrigation supports soil-friendly farming methods, and cleaner fertilizers minimize chemical runoff, protecting waterways. Reliable energy enables rural areas to grow without deforestation or harmful practices. Research in *Nature Sustainability* highlights that stable access to water and energy is essential for advancing regenerative agriculture globally.

This look at agriculture showcases the potential real-world benefits of thorium energy. From the co-production of clean water, fuel, and fertilizer to power for greenhouses and cold storage, thorium reactors can support every step of the agricultural supply chain with low-emission, local energy. This makes farming more resilient in the face of climate and resource pressures and able to prosper and thrive with more efficiency and productivity.

Thorium is Rewriting the Story of Nuclear Waste and Nuclear Safety

With so many positive environmental benefits, supporters wonder why fielding this technology has taken so long. One primary reason is that nuclear energy has a public image problem, primarily because of words like "nuclear waste" and "nuclear meltdown." But that perception is changing, and when it comes to thorium, much of the fear is based on outdated and inaccurate information.

Upcycling Nuclear Waste

Thorium offers a compelling alternative. It generates significantly less long-lived radioactive waste than uranium-based reactors, and its waste decays to background radiation levels much faster. Many of thorium's byproducts, including medical isotopes, can be extracted for beneficial use. Best of all, advanced reactor designs like molten salt reactors can be configured to consume existing stockpiles of uranium waste or even turn surplus weapons-grade plutonium into usable reactor fuel, converting a liability into clean electricity. In this way, thorium has the potential to reduce future waste and help resolve the legacy waste issues that fueled public concern in the first place.

How Thorium MSR Waste Would Be Handled

In a thorium molten salt reactor, the radioactive leftovers stay mixed in the fuel salt and can be gradually taken out using fuel salt cleanup methods like chemical treatment via fluorination, heating under pressure through vacuum distillation, or using electric currents to separate materials through electrochemical separation. This lowers the amounts of harmful leftovers that block chain reactions and helps keep the fuel working smoothly and for a longer time. Once removed, the leftover fission materials are small in volume and need less storage. Unlike traditional spent fuel rods, which remain radioactive for thousands of years and must be buried deep underground, waste from thorium molten salt reactors is easier to manage and is much less contentious. This shifts the public perception of nuclear waste from a long-term hazard to a more manageable byproduct of clean energy.

Policy Reform for Public Trust and Industry Fairness

Nuclear policy expert Jack Spencer argues that the US must move away from a government-controlled, one-size-fits-all approach to nuclear waste management. He proposes a system in which private nuclear operators are responsible for managing their own waste, using funds collected from electricity sales, but with clear legal frameworks and oversight to ensure safety and transparency. Spencer emphasizes that this model would allow competition and innovation to flourish in waste processing and storage, while giving the public confidence that waste is being handled responsibly. Most importantly, it aligns incentives so that utilities invest in advanced reactor designs, such as thorium MSRs, which inherently reduce long-term waste burdens. Spencer's model reflects a growing consensus that a transparent, decentralized, and performance-based waste policy is critical to public acceptance of nuclear energy.[12]

Freeze Plugs Instead of Meltdowns

As mentioned earlier, thorium reactors have advanced safety systems compared to older uranium designs. They don't need high-pressure water to stay cool. Instead, if something goes wrong, a special plug melts and lets the liquid fuel flow into a safe tank, automatically shutting down the reactor without human intervention. Canada and India are implementing passive safety features, including freeze plugs, to minimize accidents and enhance public confidence in thorium as a safe energy option. Canadian regulators have reviewed thorium reactor proposals and deemed them among the safest nuclear technologies available. In the US, the Department of Energy is backing early-stage thorium research, indicating strong confidence in its safety and waste management benefits.

Overcoming Barriers: Public Trust and Nuclear Perception

Thorium developers engage local communities through forums, educational campaigns, and collaborations with environmental groups to shift perceptions from "risky nuclear" to "safe, smart nuclear." This approach highlights sustainability and transparency, addressing public concerns about safety and waste. Building trust and political support requires clear communication, education, and community involvement.

Building Trust in Clean Power

Thorium is more than an energy source; it can help reduce emissions, restore ecosystems, and stabilize our environment. Public trust is vital for energy technology success, making community engagement crucial for transforming thorium from a mystery into a movement.

Building Trust, One Community at a Time

Across Canada and parts of Europe, thorium developers are taking a new approach. They are inviting communities to help shape upcoming projects. In Alberta, a pilot program has partnered with Indigenous communities[13] to plan small thorium-powered microgrids tailored to their unique needs. This initiative not only provides energy but also supports self-determination and cultural preservation. Education plays an important role, too.

In France and Japan, universities offer nuclear innovation courses[14], including hands-on work with molten salt reactor concepts.

In Norway, high school programs introduce students to nuclear science[15] through interactive workshops and online learning, making the next generation part of the solution.

On social media, thorium advocates can share progress and answer questions. These open conversations on YouTube, Reddit, and community livestreams have helped clarify misconceptions and connect people who want to be part of a cleaner energy future. And the results are showing.

In Finland, public support for nuclear energy has increased significantly[16] due to effective engagement in discussions on reactor design and waste storage. A similar global shift for thorium could occur if transparency and collaboration are prioritized.

Looking Ahead

While thorium isn't a complete solution for environmental issues, it is an effective tool for producing clean energy, safe water, abundant food, and environmental protection. Thorium reactors can contribute to a more sustainable world by lowering pollution, improving waste management, and enhancing energy access.

In light of rising energy costs and climate challenges, thorium technology can offer a clean, stable, and profitable energy solution, transforming how we power industries and agriculture. It can reduce dependence on fossil fuels and mitigate energy market risks, protecting against future uncertainties.

To enable thorium's success, governments must develop regulations that facilitate the approval and construction of new reactors tailored to thorium's design and safety features. Investments in research, infrastructure, and public education about nuclear power's advantages are also critical.

As we've seen, thorium energy holds the power to transform agriculture by making food systems cleaner, more resilient, and more equitable. But farming the Earth is only part of the story. As humanity looks beyond our home planet, the need for compact, high-output energy systems becomes even more apparent. The following chapters explore how thorium and molten salt technologies may offer a stepping stone to the long-sought dream of fusion power and how these innovations could help us thrive on the Moon, Mars, and beyond.

1. UNITED NATIONS DEPARTMENT OF ECONOMIC AND SOCIAL AFFAIRS, *WORLD POPULATION PROSPECTS 2022* (NEW YORK: UNITED NATIONS, 2022).

2. FOOD AND AGRICULTURE ORGANIZATION OF THE UNITED NATIONS, *HOW TO FEED THE WORLD IN 2050* (ROME: FAO, 2009).

3. INTERNATIONAL ATOMIC ENERGY AGENCY, *INTRODUCTION OF NUCLEAR DESALINATION: A GUIDEBOOK*, TECHNICAL REPORTS SERIES NO. 400 (VIENNA: IAEA, 2000).

4. L.M. MORTENSEN, "CO_2 ENRICHMENT IN GREENHOUSES," *AGRICULTURAL AND FOREST METEOROLOGY* 49, NO. 3 (1990): 209–224.

5. FLIBE ENERGY, "MSR FEASIBILITY AND SITING STUDY FOR AGRICULTURAL USE," INTERNAL WHITE PAPER, 2023.

6. US DEPARTMENT OF ENERGY, IDAHO NATIONAL LABORATORY, "HIGH TEMPERATURE ELECTROLYSIS FOR HYDROGEN PRODUCTION," ACCESSED 2023.

7. US DEPARTMENT OF ENERGY, ADVANCED MANUFACTURING OFFICE, *INDUSTRIAL DECARBONIZATION ROADMAP* (WASHINGTON, DC: DOE, 2021).

8. ALASKA CENTER FOR ENERGY AND POWER, "ENERGY AND FOOD RESILIENCE IN REMOTE ALASKA," 2022 BRIEFING PAPER.

9. US DEPARTMENT OF ENERGY, OFFICE OF NUCLEAR ENERGY, "ALASKA MICROREACTOR INITIATIVE," PRESS RELEASE, APRIL 2022.

10. J.A. RAKOCY ET AL., "AQUAPONIC PRODUCTION OF TILAPIA AND BASIL," *UNIVERSITY OF THE VIRGIN ISLANDS RESEARCH BULLETIN*, 2004.

11. M.E. DRY, "FISCHER–TROPSCH PROCESSES," *CATALYSIS TODAY* 71, NO. 3–4 (2002): 227–241.

12. JACK SPENCER, NUCLEAR REVOLUTION: POWERING THE NEXT GENERATION (WASHINGTON, DC: THE HERITAGE FOUNDATION PRESS, 2024).

13. THORIUM CANADA, "COMMUNITY ENGAGEMENT FOR MICROGRID SITING: ALBERTA INDIGENOUS PILOT," 2023 FIELD REPORT.

14. OECD NUCLEAR ENERGY AGENCY, *NUCLEAR EDUCATION AND TRAINING: FROM CONCERN TO CAPABILITY* (PARIS: NEA, 2021).
15. NORWEGIAN MINISTRY OF EDUCATION, "STEM ENRICHMENT PROGRAMS IN UPPER SECONDARY SCHOOLS," 2020.
16. WORLD NUCLEAR NEWS, "FINNISH PUBLIC BACKS NEW NUCLEAR," JANUARY 2023.

Part Four

Part IV: The Long Horizon

Chapter 9

Thorium Synergy in the Future of Fusion

N uclear fusion promises endless clean energy, like having a star in a bottle. However, the biggest challenge with fusion isn't just starting it; it's about keeping it going and getting usable energy on a large scale. Thorium may provide the synergy to help with this.

Fusion's Promise...and Its Obstacles

Fusion reactors, capable of mimicking the sun's energy production, promise a clean and virtually limitless supply. By fusing light atomic nuclei, these reactors create helium, a harmless byproduct, with no risk of a runaway reaction. The energy density of fusion is immense, and projects like SPARC aim to produce more energy than the reactor consumes (known as net energy gain, or Q greater than 1).

However, the path to making fusion power ready for everyday use is still tough and unclear. Achieving the necessary super hot plasma temperatures (exceeding 100 million degrees Celsius) requires advanced materials and precise magnetic confinement. Deuterium-tritium (D-T) fusion generates high-energy neutrons that damage reactor parts, and finding a sustainable way to produce tritium inside the reactor remains an unsolved challenge. Although experimental reactors like ITER and SPARC are making progress, fusion is still decades away from being widely available. It will also require major investment and new regulations.

Fusion research has a range of timelines. SPARC[1], under development by Commonwealth Fusion Systems and MIT, hopes to produce more energy than it uses by the late 2020s. ITER[2], the largest fusion project in France, aimed for its first plasma by 2025 and hopes to begin full operation by 2035. However, these are not yet electricity-generating reactors. Commercial fusion power plants may not be ready until the 2040s, with broader adoption in the 2050s. In contrast, thorium molten salt reactors (MSRs) are beginning deployment, helping provide cleaner energy during the critical years when cutting global emissions is most urgent.

Thorium MSRs: The Practical Path Forward

Thorium MSRs are an immediate solution for clean energy. They operate safely at atmospheric pressure with liquid fuel that shuts down automatically in case of overheating. While challenges like corrosion-resistant materials and outdated regulations exist, they can be addressed. Unlike fusion, thorium MSRs can be deployed relatively quickly if political and financial support align.

Fusion Reactor vs. Thorium MSR Cost and Timeline

Practical solutions often lack media attention and funding. While fusion is intriguing, thorium fission meets current energy needs. To provide clean, reliable power to billions, resources should prioritize immediacy and efficiency.

Investing in thorium MSRs presents lower risks and quicker returns, which will also strengthen supply chains and regulatory expertise necessary for future fusion developments.

THORIUM MSR VS. FUSION PROTOTYPES

FEATURE	THORIUM MOLTEN SALT REACTORS (MSRS)	SPARC (FUSION PROTOTYPE)	ITER (INTERNATIONAL FUSION REACTOR)
Technology Type	Fission using thorium in molten salt	Deuterium-tritium fusion (tokamak)	Deuterium-tritium fusion (tokamak)
Target Milestone Date	2025–2030 (pilot deployment)	Net energy gain by ~2025	First plasma: 2035
Electricity Output (Projected)	50–500 MWe (modular designs)	0 MWe (scientific gain, no power output)	0 MWe (scientific gain, no power output)
Commercial Power Plant Target	Early 2030s (modular SMRs)	~$3B for SPARC (no power demo)	No—no thermal-to-electric conversion planned
Capital Cost Estimate	$300M–$2B per plant (scalable)	~$3B for SPARC (no power demo)	No—no thermal-to-electric conversion planned
Use of Waste Heat	Near-term (~10 years), scalable with existing grid	No—plasma heat not captured for use	No—no thermal-to-electric conversion planned
Grid Integration Potential	Regulatory modernization, corrosion-resistant materials	Extreme magnetic confinement, tritium supply	Political and financial complexity, materials resilience
Strategic Advantage	Readiness for power, water, and heat	Water, and other heat applications now	Multinational proof-of-concept for plasma physics

Fusion and Fission: A Complement...Or Crossroads?

One perspective is that thorium fission and D-T fusion are part of a shared technological future, each contributing to a resilient and diverse energy ecosystem. Yet not all experts see it this way. Kirk Sorensen, founder of Flibe Energy and a pioneer of modern thorium advocacy, once a proponent of molten salts as a foundational platform for fusion systems, has grown increasingly critical of pursuing fusion itself.

In his recent commentaries, Sorensen highlights the complexity of fusion energy, which demands temperatures over 100 million degrees Celsius, specialized materials to withstand neutron bombardment, and elaborate confinement systems. "You don't need a miniature sun to boil water," he quipped, noting that thorium MSRs already produce industrial-grade heat at 600 °C. This temperature is sufficient for electricity, desalination, hydrogen synthesis, and industrial processing.[3]

His concern is not with the theoretical science of fusion but its engineering feasibility and economic wisdom. Fusion, he argues, offers "high pain for low gain."[4] The immense infrastructure and precision required to build and sustain plasma confinement systems could divert capital away from thorium technologies that are ready to scale now. He warns that investors chasing the dream of fusion may be missing the chance to back a cleaner, simpler solution hiding in plain sight.

These criticisms are not meant to derail fusion research but to encourage prioritization. The ongoing debate highlights a crucial strategic point: while fusion might someday fulfill its promise, thorium can achieve its potential today.

A Shared Technology Base

Some elements of fission and fusion have shared features that add value to both. FLiBe salt, a fluoride, lithium, and beryllium mixture, is likely the most viable material for fusion breeding blankets. In fusion reactors, it helps produce tritium from lithium and boosts neutron levels through its reaction with beryllium. Its chemical and heat stability lets it handle the intense radiation and high temperatures in fusion conditions.

Fusion and Fission Accelerated by AI

Fusion and fission reactors are complex dynamic systems, and artificial intelligence (AI) is becoming essential to manage them. As AI technology advances, it will accelerate the development of thorium fission and fusion more effectively than traditional engineering methods.

At Argonne National Laboratory, AI is speeding up the development of molten salt chemistry, reducing the time needed for experiments from years to weeks. AI can optimize salt chemistry in thorium MSRs, extending fuel cycles and enhancing safety.[5]

For fusion, AI aids in keeping the super-hot plasma stable and can predict problems before they happen, which is crucial for sustaining fusion reactions over time. It can also help adjust the chemical makeup of FLiBe blankets for breeding the tritium needed for fusion.

Fusion and Fission Share a Future

As companies like Commonwealth Fusion Systems and TAE Technologies advance, their material suppliers and reactor systems may increasingly overlap with thorium fission developers. Shared materials. Shared engineering. Shared heat loops. Eventually, shared energy markets. This points to a future where fission and fusion share a grid and perhaps lead to fission-fusion hybrid solutions that use the best of both technologies.

A Multi-Solution Reactor Grid

Energy discussions often fixate on a single solution, but the 21st century will demand a diverse network of reactors. Different reactors will serve specific purposes: some for desalination, others for medical isotope production, synthetic fuels, or fusion-breeding.

Thorium MSRs can begin creating this infrastructure, while fusion technology will develop over time. By establishing complementary systems today, we can enable both technologies to coexist and support the global energy landscape.

The future of nuclear energy isn't about selecting one ideal technology; it's about creating a flexible, resilient, and sustainable network of solutions to meet the changing needs of our century and beyond.

Toward a Hybrid Energy Future

In a few decades, we may see hybrid grids that combine the strengths of fission and fusion. Thorium MSRs could provide steady, nonstop power, while fusion reactors supply flexible, high-output energy when demand is high. Both could use the same salt-based FLiBe systems for tritium breeding and heat management. However, integrating these systems is not just a matter of plugging one into the other.

It presents real engineering and operational challenges that must be addressed if fission-fusion hybrids are to become practical:

Neutron economy management: Fusion reactions, particularly deuterium-tritium (D-T) fusion, release fast neutrons at 14.1 mega electron volts[6], far more energetic than neutrons typically produced in fission reactors. Effectively harnessing these high-energy neutrons without causing material degradation or compromising the integrity of breeding blankets demands advanced shielding technologies, neutron moderation techniques, and the development of next-generation structural alloys capable of withstanding intense neutron bombardment and prolonged radiation exposure.

Tritium containment: As a radioactive isotope of hydrogen, tritium can pass through metals and escape through microscopic flaws, which means carefully designed systems are needed to contain and collect it safely and efficiently.

Current Fission-Fusion Hybrid Experiments

The foundations for shared chemistries, AI smart control systems, and small, easy-to-build reactor designs, are being established in national labs, startups, and fusion research programs. Several design efforts offer a glimpse at how fusion-fission synergies might be achieved.

The Fusion-Fission ion Hybrid (FFH) reactor concept studied at Los Alamos National Laboratory[7] uses a fusion core to trigger safe nuclear reactions in a surrounding layer of thorium or used reactor fuel, all operating below the level needed to maintain a self-sustaining reaction.[8]

China's Fusion-Fission Hybrid Reactor (FFHR) program[9] aims to produce usable nuclear fuel from thorium by using neutrons from D-T fusion, effectively turning fusion neutrons into a trigger for a liquid-fuel molten salt fission reaction.[10]

Hypothetical Applications of Fission-Fusion Hybrids

Future visionary and hypothetical implementations, if innovators take up the challenge, could include:

Remote coastal hybrid facilities which combine a thorium molten salt reactor to provide steady heat and power with a fusion unit that meets peak electricity needs and produces tritium on-site using a special salt mixture (FLiBe). This setup could also support a large-scale reverse osmosis (RO) salt water purification system, delivering energy, fresh water, and artificial hydrogen from seawater, all within a single, integrated plant.

Beyond Earth, this hybrid vision expands:

A lunar base, for instance, could use thorium molten salt reactors to provide steady and reliable baseline power, while small fusion reactors supply extra energy during peak demands like processing lunar regolith soil or supporting life-support system surges. Both reactors might share a common FLiBe system to produce tritium and transfer heat, improving reliability and overall efficiency.

Perhaps most visionary is the prospect of space travel:

A deep space craft using a fission-fusion hybrid for propulsion and onboard systems. The fission unit would power long-duration life support, communications, and navigation over long missions, while the fusion drive would deliver powerful bursts of thrust for fast travel across vast distances. By pairing both technologies, such a vessel could dramatically extend humanity's reach across the solar system while minimizing the energy payload required from Earth.

As hybrid concepts mature, they remind us that energy innovation need not be a zero-sum game. When the strengths of different nuclear technologies are harmonized, the result is more than additive; it becomes transformational.

In the next chapter, we'll explore how thorium's compact, high-reliability energy potential positions it as a key enabler for off-world settlements, deep space travel, and a sustainable human presence beyond Earth.

Endnotes — Chapter 9

1. COMMONWEALTH FUSION SYSTEMS. "SPARC: A COMPACT, HIGH-FIELD, NET ENERGY FUSION DEVICE." ACCESSED MAY 2025. HTTPS://CFS.ENERGY/SPARC/
2. ITER ORGANIZATION. "THE ITER PROJECT." ACCESSED MAY 2025. HTTPS://WWW.ITER.ORG/PROJ/INAFEWLINES
3. KIRK SORENSEN. "FUSION VERSUS THORIUM." PRESENTATION AT TEAC 10, THORIUM ENERGY ALLIANCE, OCTOBER 2023. HTTPS://WWW.YOUTUBE.COM/WATCH?V=VIDEO_ID

4. Kirk Sorensen. "High Pain for Low Gain: Why Fusion Isn't the Answer." Energyfromthorium.com. Accessed May 2025. Https://energyfromthorium.com/fusion-critique/

5. Moisseytsev, A., and T. J. Heidet. "Molten Salt Reactor Design Innovations Enabled by AI." Argonne National Laboratory, 2024. Https://www.anl.gov/article/molten-salt-design-with-ai

6. Ignatiev, V., et al. "Development of Flibe for Use in Fusion-Fission Hybrids." Fusion Engineering and Design, vol. 136, 2018, pp. 1101–1107.

7. Los Alamos National Laboratory. "Fusion-Fission Hybrid Systems for Actinide Management." La-ur-14-27624, 2014.

8. J. D. Sethian et al., "The Fusion-Fission Hybrid (FFH) Reactor Concept for Advanced Fuel Cycles," Los Alamos National Laboratory Technical Report La-ur-10-05645, 2010. Https://permalink.lanl.gov/object/tr?what=info:lanl-repo/lareport/la-ur-10-05645

9. Institute of Plasma Physics, Chinese Academy of Sciences. "Fusion-Fission Hybrid Reactor Conceptual Design Report." Hefei, China, 2022.

10. Jiangang Li et al., "Fusion–Fission Hybrid Reactor (FFHR) Research Activities in China," Nuclear Fusion 59, no. 11 (2019): 112003. Https://doi.org/10.1088/1741-4326/ab3c61

1. COMMONWEALTH FUSION SYSTEMS. "SPARC: A COMPACT, HIGH-FIELD, NET ENERGY FUSION DEVICE." ACCESSED MAY 2025. HTTPS://CFS.ENERGY/SPARC/

2. ITER ORGANIZATION. "THE ITER PROJECT." ACCESSED MAY 2025. HTTPS://WWW.ITER.ORG/PROJ/INAFEWLINES

3. KIRK SORENSEN. "FUSION VERSUS THORIUM." PRESENTATION AT TEAC10, THORIUM ENERGY ALLIANCE, OCTOBER 2023. HTTPS://WWW.YOUTUBE.COM/WATCH?V=VIDEO_ID

4. KIRK SORENSEN. "HIGH PAIN FOR LOW GAIN: WHY FUSION ISN'T THE ANSWER." ENERGYFROMTHORIUM.COM. ACCESSED MAY 2025. HTTPS://ENERGYFROMTHORIUM.COM/FUSION-CRITIQUE/

5. MOISSEYTSEV, A., AND T. J. HEIDET. "MOLTEN SALT REACTOR DESIGN INNOVATIONS ENABLED BY AI." ARGONNE NATIONAL LABORATORY, 2024. HTTPS://WWW.ANL.GOV/ARTICLE/MOLTEN-SALT-DESIGN-WITH-AI

6. IGNATIEV, V., ET AL. "DEVELOPMENT OF FLIBE FOR USE IN FUSION-FISSION HYBRIDS." FUSION ENGINEERING AND DESIGN, VOL. 136, 2018, PP. 1101–1107.

7. LOS ALAMOS NATIONAL LABORATORY. "FUSION-FISSION HYBRID SYSTEMS FOR ACTINIDE MANAGEMENT." LA-UR-14-27624, 2014.

8. J. D. SETHIAN ET AL., "THE FUSION-FISSION HYBRID (FFH) REACTOR CONCEPT FOR ADVANCED FUEL CYCLES," LOS ALAMOS NATIONAL LABORATORY TECHNICAL REPORT LA-UR-10-05645, 2010. HTTPS://PERMALINK.LANL.GOV/OBJECT/TR?WHAT=INFO:LANL-REPO/LAREPORT/LA-UR-10-05645

9. INSTITUTE OF PLASMA PHYSICS, CHINESE ACADEMY OF SCIENCES. "FUSION-FISSION HYBRID REACTOR CONCEPTUAL DESIGN REPORT." HEFEI, CHINA, 2022.

10. JIANGANG LI ET AL., "FUSION–FISSION HYBRID REACTOR (FFHR) RESEARCH ACTIVITIES IN CHINA," NUCLEAR FUSION 59, NO. 11 (2019): 112003. HTTPS://DOI.ORG/10.1088/1741-4326/AB3C61

Chapter 10

Powering the Final Frontier: Thorium MSRs and the Architecture of Space Civilization

The Challenge

Space invites wonder, but it also challenges our presence. It is a realm of silence and radiation, filled with dazzling stars and perilous vacuums. To survive in space, energy is vital. There are no fossil fuels, no backup grids, and no margin for error. Whether you are in a lunar habitat, exploring the surface of Mars, or traveling between planets, power is more than a convenience; it is essential for life.

Solar panels can be effective near Earth, but they cannot provide power during the two-week lunar night, leading to drained batteries. Traditional nuclear power sources, such as radioisotope generators, produce insufficient energy for growing settlements.

So, what energy source could enable sustainable living in deep space?

This is why thorium is needed. Thorium-fueled molten salt reactors are compact, reliable, and long-lasting, providing a consistent power supply in environments where sunlight is nonexistent and survival is at stake. With adequate heat and electricity from these reactors, we could operate life support systems, cultivate crops, purify water, construct habitats, and even launch expeditions to other planets.

In 2025, the US government officially recognized space-based nuclear power as a national priority. As we look ahead to the future of space exploration, these advancements in energy technology could transform our ability to thrive beyond Earth.

What follows is a step-by-step roadmap for bold space exploration, envisioning how thorium-fueled solutions could enable our future progress in space.

Space Thorium MSR Roadmap: Step by Step Development Framework

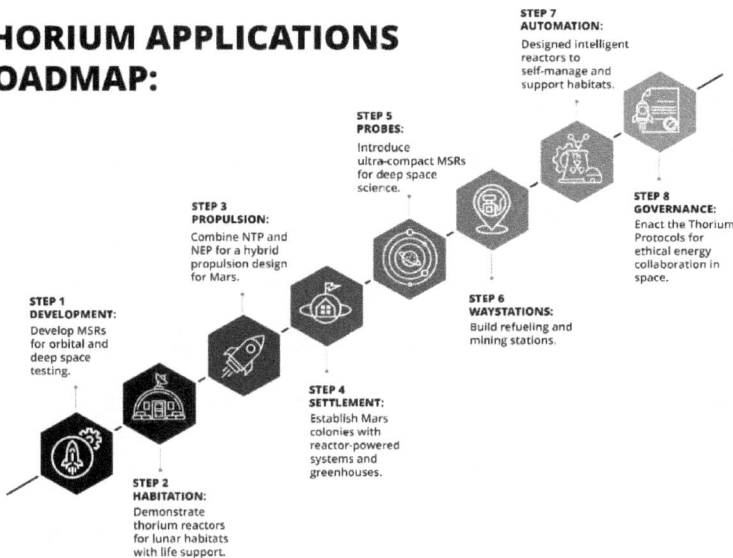

THORIUM APPLICATIONS ROADMAP:

STEP 1 DEVELOPMENT: Develop MSRs for orbital and deep space testing.

STEP 2 HABITATION: Demonstrate thorium reactors for lunar habitats with life support.

STEP 3 PROPULSION: Combine NTP and NEP for a hybrid propulsion design for Mars.

STEP 4 SETTLEMENT: Establish Mars colonies with reactor-powered systems and greenhouses.

STEP 5 PROBES: Introduce ultra-compact MSRs for deep space science.

STEP 6 WAYSTATIONS: Build refueling and mining stations.

STEP 7 AUTOMATION: Designed intelligent reactors to self-manage and support habitats.

STEP 8 GOVERNANCE: Enact the Thorium Protocols for ethical energy collaboration in space.

Step 1: Development: Proving Power in the Void

Space is not a forgiving place to test your first systems. It's cold. It's empty. And above all, it's unresponsive to your problems. If something fails out there, no technician comes to fix it. That's why the first step toward a spacefaring energy architecture begins with validation: testing thorium MSRs in orbit and deep space conditions.

We start with a self-contained reactor, autonomous, steady, and positioned into orbit or riding aboard a lander. Inside it, a molten salt reactor comes to life. Not in a lab. Not under supervision. But alone, in zero gravity.

This isn't just about proving that MSRs work. That's already been done on Earth. The challenge is proving that they work in space: that they can manage heat without air, regulate themselves without intervention, and operate safely without a single moving part depending on gravity.

System-Level Capabilities Being Demonstrated

Heat rejection through radiative panels engineered for space: since heat can't be lost through convection, it must be emitted as infrared radiation. That requires high-emissivity radiators designed to operate in vacuum, perhaps with sodium heat pipes or molten metal loops to distribute thermal loads.

Power conversion using low-maintenance cycles: Stirling engines and closed-loop Brayton systems that convert thermal energy into electricity, scaled to the 10–40 kWe class for initial orbital applications.

Autonomous reactor control: real-time diagnostics, thermal regulation, and startup/shutdown protocols, managed by embedded AI. This is where flight avionics meets reactor engineering.

Modular shielding: these are lightweight solutions that focus radiation away from sensitive equipment, possibly using water tanks, special heat-absorbing materials, or fold-out panels that cast a protective "shadow," blocking radiation where protection is needed.

This first wave of missions will power small stations, deep-space telescopes, or long-endurance satellites. But more importantly, they'll prove that reactors don't have to be watched constantly to work. They can operate in silence, through the long night of space, supplying continuous energy to instruments, habitats, or propulsion systems. These first reactors are like scouts. They go ahead, operate in new environments, and tell us what is possible.

Proving power in space involves more than just managing heat and electricity; there's another, less visible, yet equally unforgiving challenge: radiation. Beyond Earth's protective shield, space is filled with high-energy particles. Some of these particles originate from the Sun, particularly during solar storms, when sudden bursts of radiation are hurled into space. Others come from distant galaxies in the form of cosmic rays traveling near the speed of light. These high-energy particles can penetrate metals, corrupt data, flip bits in memory, disrupt navigation systems, and gradually degrade the materials we rely on.

A space-based reactor must contend with radiation from multiple sources. It generates its own neutron and gamma radiation while also facing constant bombardment from cosmic rays. This dual challenge means that the solutions implemented need to be exceptionally robust. As a result, protecting systems from radiation is a key part of the design, starting with the very first test flights.

What does radiation protection mean in practical terms?

It begins with protective barriers. Neutron radiation, produced within the reactor, is absorbed using specialized materials such as boron-coated tanks, hydrogen-rich compounds, and thick protective layers. Gamma rays are more difficult to block and require heavy materials to stop them effectively. These protective layers can be shaped or even built from local materials, such as moon dust. In some designs, water or fuel tanks can double as protective shields, serving their main purpose while quietly protecting against radiation

Outside the spacecraft, we face the problem of cosmic rays and solar particles that are constantly present. To survive these, onboard electronics must be built to resist radiation, meaning they need to be equipped with fault-tolerant chips, backup systems, and the capability to reboot, reset, or reroute power in the event of a malfunction.

Even the reactor's control systems, powered by artificial intelligence, must be designed to handle radiation. These are not simple thermostats; they are smart learning systems programmed to detect heat spikes, magnetic issues, or minor malfunctions before they get worse. Most importantly, they must operate without human intervention.

Moreover, the materials that constitute the reactor's structure, such as pipes, tanks, sensors, and cooling panels, must withstand years of combined radiation exposure without breaking, swelling, or losing functionality. This requires careful testing, the use of specialized metals, and multiple layers of protection.

In summary, protecting against radiation involves more than just covering a single part. It's about ensuring that the entire system, from fuel salt to final output, continues to function effectively for years in space, one of the harshest environments known to humanity. The initial reactors will evaluate their power generation capability and long-term durability. They must effectively manage heat, maintain control, handle waste, and resist internal and external radiation to establish a solid foundation for future developments.

Step 2: The Moon Colony: Establishing a Foothold

The Moon is closer than Mars. We've already walked there. But no one has ever lived there.

To survive the two-week lunar night, when the sun vanishes behind the horizon and temperatures drop to -250°C, we need something better than solar panels and batteries. We need a source of heat that never stops. A power system that endures the cold, ignores the darkness, and doesn't care whether the weather is cooperating, because on the Moon, there is no weather.

Imagine a dome, settled in the shadow of a southern crater, where ice stays hidden in constant low light. Under a meter of insulating moon dust, a reactor power unit runs quietly. Heat flows through pipes. Lights shine inside the shelter. A greenhouse is full of life under bright LED lights. The air is cleaned, water is reused, and plants grow while the land outside stays frozen and quiet. This is no longer exploration. This is habitation.

Moon Base Powered by Thorium Molten Salt Reactors

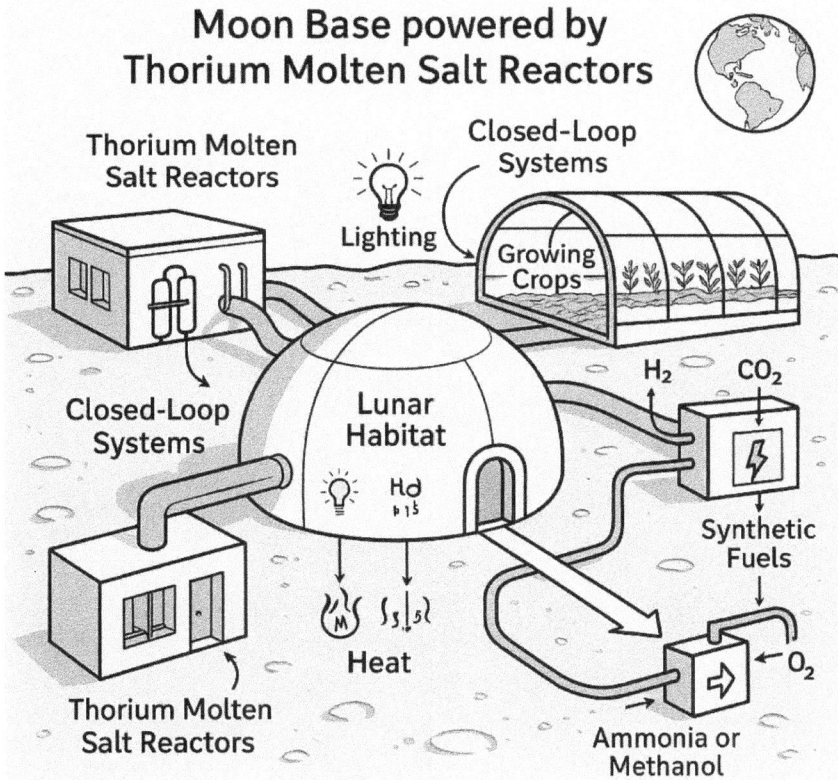

Moon Base powered by
Thorium Molten Salt Reactors

Thorium Molten Salt Reactors

Closed-Loop Systems

Lighting

Growing Crops

Closed-Loop Systems

Lunar Habitat

H_2 CO_2

Synthetic Fuels

Thorium Molten Salt Reactors

Heat

Ammonia or Methanol

O_2

System-Level Capabilities Enabled by the MSR

A 40–100 kWe molten salt reactor: Designed for long-duration, low-maintenance lunar operation. Buried for protection from radiation and to hold in heat, its energy footprint supports everything a small lunar colony needs.

Thermal-electric co-generation: High-grade electricity runs life support, communications, and tools; low-grade waste heat is used to warm greenhouses, help recycle water, and stop systems from freezing.

Closed-loop life support, powered by reactor heat: Oxygen is made by splitting water with electricity, water is recovered from the air, and carbon dioxide is removed to keep the air safe to breathe

Robotic integration: Regolith refers to the dusty and rocky material found on the surface of the moon, and it can be put to practical use. Energy from a reactor powers machines that process this regolith, allowing for the separation of valuable elements, such as naturally occurring thorium. On-site manufacturing units can convert regolith into useful items, such as bricks for building structures. Additionally, robotics can operate 3D printers to create and repair parts as needed.

Energy storage balancing: Heat can be stored in specialized phase-change materials designed to capture energy, or used to warm systems during the cold lunar nights. This helps balance power consumption and prolongs the lifespan of equipment.

Thorium molten salt reactors are particularly well-suited for this purpose. They operate at atmospheric pressure, which means they do not rely on bulky, high-pressure tanks. These reactors generate both electricity and heat with high efficiency, can operate for years without needing to refuel, and can refuel while operating.

For the first time, humans would live entirely off Earth. They would not depend on regular shipments from the home planet but instead rely on a self-sustaining system. Every aspect, from recycling and agriculture to power generation and shelter, is built around one crucial component: a reactor that never shuts down.

This lunar colony serves as a blueprint for future endeavors. If we can thrive here, in the silent cold of the Moon, then everything from Mars to the asteroid belt is within our reach.

Step 3: The Journey to Mars: Propulsion for the Next Frontier

Reaching Mars involves more than just arriving; it requires getting there quickly, with everything needed for the journey, and landing safely while remaining operational. Traditional chemical rockets alone are inadequate. They burn intensely for a short time then coast for months, leaving the spacecraft with little control over where it goes. To address this challenge, we need engines that work in stages by providing short, powerful pushes to escape gravity, then providing steady, long-term acceleration through deep space. Nuclear propulsion is the solution.

Thorium reactors offer a unique advantage by supporting both nuclear thermal propulsion (NTP) and nuclear electric propulsion (NEP) within the same framework. This combination provides both high thrust and continuous motion.

System-Level Capabilities in the Mars Transit Vehicle

Dual-Mode Reactor Operation: A thorium molten salt reactor delivers continuous heat for electric systems and can also produce quick bursts of heat for strong engine pushes. Switching between these modes is handled automatically.

Brayton Cycle Electric Conversion: This is a power system that uses heat from a reactor, such as a thorium molten salt reactor, to spin a turbine and generate electricity, much like a high-efficiency jet engine. That electricity can then be used to power ion drives, which are propulsion systems that work very differently from traditional rockets. Instead of burning fuel, ion drives use electricity to charge and accelerate inert gases, usually xenon, out the back of the spacecraft. The force is gentle but constant, allowing the ship to build up speed over long periods, making them ideal for deep space missions, where efficiency and endurance matter. Plus inert gases like xenon are safe, stable, and easy to store.

Ion propulsion is not theoretical; it's a proven technology. NASA has already used it on missions like Dawn and Deep Space 1, and it has been flying for decades on certain commercial satellites. What's new is pairing these engines with compact nuclear power sources, such as a thorium reactor in a Brayton system, which can sustain a much higher energy output than solar panels alone. At 150 to 250 kilowatts, this setup enables faster travel and heavier payloads for deep space missions, enabling a crewed exploration of Mars or other sustained interplanetary travel and long-range science missions.

Hydrogen Thermal Propulsion: For bursts of thrust, we can use heat from the thorium reactor to directly heat hydrogen gas. As the hydrogen is superheated, it rapidly expands and is expelled through a nozzle to produce thrust. This method delivers powerful, short bursts of acceleration, making it especially useful for critical maneuvers like entering orbit, changing trajectory, or escaping a planet's gravity.

Thermal Management Through Deployable Radiators: Wide cooling fins help control the reactor's heat output, ensuring the system remains stable and intact during long distance travel.

Artificial Gravity via Rotation: With excess power, the ship utilizes mechanical systems to simulate Mars-level gravity, thereby helping to mitigate muscle loss and fluid shifts during the journey.

Closed-Loop Habitat Support: This includes splitting water to make oxygen, removing carbon dioxide from the air, and recycling water, all powered by the reactor's electricity.

This spacecraft functions as a moving city. It cultivates food, recycles materials, supports science labs, command modules, and medical bays. Importantly, it does not coast; it accelerates, decelerates, and ensures survival during re-entry.

With thorium as the main power source, the engine does more than just push the ship, it keeps it moving steadily, allows for course changes, and powers all systems without stopping during the trip. The journey to Mars is a test of distance and energy continuity. With thorium molten salt reactors, we are well-equipped to pass this challenge.

Step 4: Mars: Energy for a Second Home

Living on Mars demands complete independence. Running a Mars colony cannot rely on fuel convoys from Earth. While solar panels may provide energy during daylight, Mars dust storms can last for weeks, and no help can arrive during those blackouts.

What is needed is a permanent energy backbone. On Earth, we refer to this as a power grid. On Mars, we will have to build one from scratch, with two liquid salt reactors at its core.

System-Level Infrastructure at the Martian Settlement

A twin-reactor hub, with each reactor producing 60–80 kWe, the system offers backup power, balanced energy sharing, and full energy independence from start to finish.

In-situ resource utilization (ISRU): In-situ resource utilization is the process of using local materials on Mars to produce what astronauts need—like air, water, and fuel—to sustain human missions and support long-term settlements without depending entirely on costly shipments from Earth. One of the biggest opportunities lies in Mars' atmosphere, which is rich in carbon dioxide. Using electrolysis, we can split CO_2 to produce oxygen for breathing and combine it with hydrogen to create methane fuel: ideal for return trips or powering surface vehicles. But to make that happen, we also need water. And while Mars may look dry on the surface, we know from satellite scans and lander data that there's frozen water beneath the soil, especially near the poles and at certain mid-latitudes. Extracting that water means melting and purifying ice from the Martian regolith. That's where reliable heat and power are essential. Thorium-based microreactors could supply the steady energy needed to power those electrolysis units. They'd provide both the electricity to split molecules and the thermal energy to warm and process subsurface ice, turning it into liquid water ready for drinking, plant growth, or fuel production.

High-temperature regolith processing: Furnaces powered by the heat from the MSRs transform Martian soil into ceramic bricks, tiles, and other building materials.

Thermal-linked greenhouse systems: Waste heat from the reactors will keep agricultural domes warm during freezing nights, balance moisture levels, and keep plant lights running nonstop.

Radiation shielding via regolith sintering: Robots will heat layers of Mars soil into bricks for strong protective walls, helping block radiation and making it easier to expand the base.

Robotic fabrication powered by excess capacity: With a consistent energy supply, self-running machines can manufacture spare parts, take care of living areas, and extend infrastructure without needing supplies from Earth.

On Mars, the margin for error is slim. However, thorium reactors change the equation. They eliminate the need for a continuous fuel supply, reduce dependence on launches from Earth, and create conditions for permanent habitation. By combining heat, electricity, and fuel production into one self-sustaining system, thorium molten salt reactors not only support life on Mars but also make it sustainable long-term.

Step 5: Deep Space Probes: Power Beyond Sunlight

As we journey farther from Earth, the sun's brightness diminishes. Beyond Mars, solar panels become less effective, and past Jupiter, they are no longer practical. However, deep space offers intriguing mysteries, such as ice-covered oceans on Europa, organic lakes on Titan, and the dark edge of the Kuiper Belt.

To explore these wonders, space probes require power systems that can last for decades. This could be solved by an ultra-compact thorium microreactor, which is no larger than a barrel but can consistently provide 10 to 20 kWe for 30 to 50 years. These microreactors can act as lifelines for scientific missions in the extreme cold of deep space.

Thorium may also hold the key to the most accurate clocks ever built: clocks so stable they could guide deep-space probes across billions of miles with unmatched precision. In 2025, DARPA launched a research effort called SUNSPOT to explore this possibility. The idea is to build a new kind of timekeeper based not on electrons, like today's atomic clocks, but on the nucleus of a thorium atom. Scientists believe thorium-229 has a unique property: a nuclear "heartbeat" that ticks with astonishing regularity when hit with ultraviolet light. If researchers can tune a laser to this faint signal, they could create a tiny nuclear clock that stays accurate for decades, ideal for space probes, secure communications, and even GPS systems on the Moon or Mars. It's one more way thorium could quietly shape the future of space exploration; not just by powering our journey, but by keeping it perfectly on time.

System-Level Technologies for Long-Duration Probes

Miniaturized molten salt reactor cores are built for heat stability and safely sealed during long missions lasting decades.

Passive radiators with dust-hardened coatings, capable of withstanding freeze-thaw cycles and micro-meteor impacts without human maintenance.

Stirling or Brayton conversion systems are designed for small footprints to power high-resolution sensors, radar, and ion propulsion units.

AI diagnostics and control systems are designed to operate independently for months or years, adjusting output based on mission needs.

Quantum or optical communication relays, sustained by the reactor's electrical output, allow high-bandwidth data streaming over billions of kilometers.

Where radioisotope thermoelectric generators (RTGs) diminish over time, thorium MSRs provide steady power. This means longer missions, more data collection, and new types of exploration, such as subsurface radar on Enceladus, deep drilling on Ganymede, or atmosphere sampling near exoplanet candidates.

These probes serve as more than scouts; they are the front line of our intelligence network in space and point us to the resources we need to push further.

Step 6: Mining and Refueling: Infrastructure Among the Asteroids

A civilization doesn't expand by transporting its entire supply chain with it. Instead, it builds outposts, develops supply nodes, and adapts locally.

The asteroid belt presents an opportunity to harvest materials in space and build what we need there rather than shipping everything from Earth. Imagine a mining station on a carbon-rich asteroid that drills into frozen volatiles, cracks CO_2 into fuel, and smelts regolith into usable metal. Powering all this quietly and continuously is a thorium molten salt reactor.

Sidebar: What Antarctica Taught Us About Remote Nuclear Power

Long before thorium reactors were being discussed for Mars bases or lunar habitats or asteroid mining stations, the U.S. Navy tested a compact nuclear reactor in one of the harshest places on Earth: Antarctica. In 1962, the PM-3A reactor was installed at McMurdo Station to supply power and steam heating, including for water desalination. It was not a thorium reactor – it used enriched uranium and was one of the earliest examples of a remote, self-contained small modular reactor (SMR), capable of producing 1.8 MW of electricity. But after just a decade of operation, the cracks and leaks in this pressurized water reactor led to its shutdown in 1972. The reactor and all contaminated material had to be dismantled and shipped off the continent.

The lessons were clear: remote nuclear power is technically possible, but only if the system is simple, safe, and requires minimal oversight. That's where thorium-fueled molten salt reactors have advantages. With passive safety, low pressure, and chemical stability, thorium MSRs offer the promise of delivering clean, resilient energy without the complications that plagued early designs like PM-3A.

System-Level Capabilities for In-Situ Resource Utilization Mining Platforms

Modular thorium MSRs ranging from 20 to 200 kWe, depending on the station's scale, with heat routed to life support, factories, and engine systems.

CO_2 and H_2O cracking reactors linked to the reactor's heat output to maximize efficiency when making fuel and producing oxygen.

Cryogenic storage systems are used to keep super-cold fuels like liquid methane and liquid oxygen stable during space missions. These fuels need to stay at extremely low temperatures to remain in liquid form, otherwise, they'll boil off and be lost.

To manage this, engineers use a combination of passive insulation, like a giant thermos, and smart use of surplus heat from the reactor to regulate system functions without extra energy demands. By integrating these cryogenic tanks directly into the outer shell of a thorium reactor housing, the design becomes more efficient. It saves weight, reduces complexity, and keeps the fuels protected and ready for use. Whether it's for propulsion, life support, or power generation, this approach makes the entire system more compact, stable, and mission-ready.

Robotic excavation platforms powered via tethered or wireless electric connections, capable of autonomous mining, maintenance, and repair.

On-site 3D printing systems supported by a consistent power supply, enabling the fabrication of tools, spare parts, or even new reactor modules using asteroid metals.

These waystations form the logistical backbone for trade and travel between planets. They reduce costs, enable refueling, and make missions to Mars, the outer planets, or even the return of spacecraft more feasible. Each node powered by thorium serves as a stepping stone, making deep space more accessible.

Step 7: Autonomy: Systems That Think, Repair, and Adapt

In deep space, independence is required. As missions stretch into years and distances grow beyond real-time communication, reactors and their support systems must become more than machines. They must be intelligent, self-regulating platforms that sense, respond, and repair themselves without waiting for instructions.

System-Level Autonomy Enabled by Thorium MSRs

AI-driven control architecture, capable of balancing neutron levels, adjusting power output, and triggering safety measures autonomously.

Predictive diagnostics, using sensor arrays and machine learning programs to detect damage, rust, or blockages before failure occurs.

Robotic manipulators and repair drones, using energy from the reactor, the system can take care of cooling parts, change filters, or even make and replace small parts as needed.

Autonomous life support balancing, in habitats where power, air composition, and temperature are managed together, not separately.

Programmable Power: What if we could tell radioactive materials when to release their energy by speeding them up or slowing them down as needed? That's the big idea behind DARPA's Decay on Demand program. Normally, when something is radioactive, it decays on its own schedule: sometimes over days, sometimes over thousands of years. But DARPA is exploring whether we can control that decay, using beams of energy to trigger or adjust it on command. Why does this matter for space? Because if it works, it could change everything about how we power long-term missions. Instead of hauling heavy nuclear reactors that constantly give off heat and radiation, future space travelers might carry small, quiet fuel sources that only activate when needed, like flipping on a light switch. It could also help manage nuclear waste by making it decay faster and become safer sooner. This kind of smart control adds a whole new level of flexibility to reactors. It means a system could not only generate power but adjust its fuel behavior based on mission demands, emergencies, or even habitat conditions. In the future, energy in space won't just be powerful; it could be programmable.

Integrated communications, with secure, reliable connections to Earth and the ability to keep working even during signal loss or space weather.

This approach goes beyond smart engineering; it's the foundation for running a civilization that stretches across multiple planets. Self-managing systems support human crews and boost overall performance. Thorium reactors, in particular, offer the steady and reliable energy that AI systems need to work at their full potential.

Step 8: Peace Through Power: A Shared Infrastructure for a Shared Future

Throughout history, access to energy has influenced geopolitics. The entities that control power often dominate people, territory, and trade. In the realm of space exploration, we have an opportunity to change that dynamic.

Thorium liquid salt reactors offer local and steady power in areas where fuel transportation poses risks and consistent solar access is unavailable. This innovation fosters a more balanced energy system for everyone.

Proposed Thorium Protocols for Cooperative Energy Governance

The Artemis Accords is a treaty that set an important foundation. Signed by many nations, they offer the first modern rules of the road for space: share your data, don't claim land, resolve disputes peacefully, and act transparently. But they don't yet cover one of the most powerful and potentially divisive resources in space: nuclear energy.

Once we begin building permanent habitats and industrial platforms beyond Earth, power becomes everything. It runs life support. Enables mining. Drives manufacturing. Propels ships. And unlike water or oxygen, power in space can be produced and controlled entirely by infrastructure.

Without a shared framework, energy could become a source of conflict, just as oil, rivers, and electricity grids have triggered disputes here on Earth. Below is a proposal that we can refer to as the Thorium Protocols. It's not a treaty yet. But it's an idea. A proposed extension of the Artemis principles into the nuclear age of space exploration. And if done right, it could be as important as any technology we launch.

The Thorium Protocols would be a set of international agreements designed to govern the peaceful use of space-based reactors. Not just thorium, but any fission system deployed off Earth. Its goal is simple: to ensure that nuclear energy enables cooperation, not control.

Proposed Thorium Protocols:

1. **Establishing inspection and transparency standards**: Reactors placed on the Moon, Mars, asteroids, or during space travel would need to disclose essential information, such as reactor size, location, and how they're running, allowing for transparency and preventing reactors from being hidden behind science labs or quietly turned into military tools.

2. **Supporting emergency energy-sharing agreements**: If one station faces a power failure, another, regardless of national affiliation, should be able to provide energy. This could involve connections through modular grids, docking with reactor-powered supply vehicles, or transferring stored energy from standby battery farms. In the harsh environment of space, energy should not be used as a bargaining chip when lives are at stake.

3. **Limiting territorial energy monopolies**: No base should be able to claim exclusive rights to reactors over essential resources, like the Moon's poles, regolith zones rich in thorium and minerals, or asteroid mining corridors just because they arrived first. The goal isn't to block competition, but to prevent artificial shortages in places where resources are actually abundant.

4. **Providing a conflict-resolution framework**: Disputes regarding energy access, system connections, or radiation safety should not lead to serious conflicts. Instead, they should be handled through fair negotiation using shared safety rules and technical guidelines everyone agrees on.

The Thorium Protocols would shape both policy and behavior. It would make energy access predictable, grid-compatible, and fair. It would allow bases from different countries and, eventually, different planet colonies, to plug into a shared future, instead of building in silos.

Because if we learn anything from Earth, it's that the real threats don't always come from outside. Sometimes, they come from competition over what should be common ground.

Thorium doesn't just power habitats. Or ships. Or greenhouses. It powers possibility. And if we set the rules right, it can also power peace. The Thorium Protocols, modeled on the Artemis Accords framework, could govern nuclear energy deployment beyond Earth.

THE THORIUM PROTOCOLS: A TRANSFORMATIVE ADDITION

To extend the spirit of the Artemis Accords into the nuclear age of space exploration, we propose the establishment of the Thorium Protocols as a hypothetical but vital framework that would:

Establish reactor transparency and inspection regimes, ensuring ail thorium MSRs operate under shared safety and environmental standards.

Enable emergency power-sharing agreements, allowing any base in crisis to access wattage from neighboring reactors, much like a shared electrical grid.

Set limits on territorial reactor claims, preventing any single nation or corporation from hoarding energy access in key resource zones.

Create conflict resolution mechanisms based on energy arbitration, where disputes are negotiated through energy-sharing agreements rather than territorial posturing.

It would protect against monopolies. Prevent militarization of power systems. And enable every base, no matter who built it, to thrive without hoarding energy. Thorium doesn't just power systems. It powers trust. And in space, trust is the rarest and most valuable resource of all.

Summary: Powering the Final Frontier

This chapter explored how thorium liquid salt reactors could become the key technology behind a sustainable and cooperative space-based future. Whether it's powering bases on the Moon or Mars, supporting mining stations on asteroids, or fueling deep space missions, thorium reactors offer steady, flexible, and self-sufficient energy, something that solar panels, chemical fuels, and traditional power sources can't consistently provide.

We followed the journey of thorium liquid salt reactors from their early design ideas to how they are being adapted for use in space, facing challenges like working in a vacuum, handling heat without air, operating in low gravity, and running for long periods without human help. We walked through a step-by-step vision of what thorium-fueled space exploration could look like. For a deeper dive, see the appendix for a technology companion that details the technical requirements and enabling technologies that could bring this vision to life.

Roadmap Recap:

Step 1. Development: Develop MSRs for orbital and deep space testing.

Step 2. Habitation: Demonstrate thorium reactors for lunar habitats with life support.

Step 3. Propulsion: Combine NTP and NEP for a hybrid propulsion design for Mars.

Step 4. Settlement: Establish Mars colonies with reactor-powered systems and greenhouses.

Step 5. Probes: Introduce ultra-compact MSRs for deep space science.

Step 6. Waystations: Build refueling and mining stations.

Step 7. Automation: Design intelligent reactors to self-manage and support habitats.

Step 8. Governance: Enact the Thorium Protocols for ethical energy collaboration in space.

The thorium MSR represents a foundational platform for achieving energy independence, advancing scientific knowledge, ensuring human safety, and promoting geopolitical stability. By developing these systems, we are laying the groundwork for a future of peace, prosperity, and purpose in a multi-planetary world.

The Great Convergence: A Vision Grounded in Physics

What unites the visionary ideas in this chapter is physics. Though the scope is vast, every component stems from established principles, emerging prototypes, or active engineering roadmaps.

Some technologies are already operational, including:

- Molten salt reactor chemistry and material science, proven in terrestrial labs and testbeds
- Kilopower's successful demonstration of compact, low-power space fission units
- AI-driven sensors for fault detection and predictive maintenance, already in use on Earth and in space avionics

Others are in near-term development, including:

- Lunar greenhouses and closed-loop life support, advancing through Artemis-era test programs
- Martian regolith sintering for construction, demonstrated in analog environments and robotic studies

- High-efficiency radiators and Stirling or Brayton-cycle systems, which are under prototyping in programs like NASA FSP and Copenhagen Atomic's containerized reactors.

Some concepts remain speculative but technically plausible, such as:

- Cryostasis systems for long-duration human travel
- Terraforming principles, like atmospheric thickening or greenhouse warming
- Quantum communication networks, which are emerging from early terrestrial quantum encryption platforms

Thorium serves as the unifying enabler across this spectrum. Thorium brings everything together. It's a fuel with the right qualities to deliver sustained, safe energy in the environments where we need it most. Whether it's powering a spacecraft beyond Neptune, a base embedded in the Moon's surface, or an automated drill on an asteroid, thorium reactors turn bold ideas into real possibilities, and give power to our dreams.

Powering Our Future

Power defines what we can achieve. On Earth, we've learned to utilize fire and electricity to build our civilization. In space, our capabilities will depend on our ability to generate resources in the environments we encounter. Thorium molten salt reactors offer a new type of energy source. They can provide propulsion and sustain life, outlasting fuel deliveries, outperforming solar in obscured conditions, and working in places where other energy sources fail.

Building a spacefaring civilization requires more than advanced technology; it needs trust, foresight, and collaboration. The energy abundance and independence we envision depend on both technology and a strong set of values. We must avoid repeating the mistakes of scarcity-driven politics in space. The Thorium Protocols and similar frameworks can help us manage conflicts before they occur, making energy a tool for shared growth rather than a source of tension.

Our goal should be to transform new worlds into habitable places and mature from merely exploring them to truly deserving to inhabit them. What kind of future will we fuel with thorium? The final chapter challenges innovators, investors, policymakers, and you.

Conclusion: A Future Worth Building

The first step of innovation is to ask ourselves the big "what if" questions. In this book, we have posed major "what if" questions, and the answers we've outlined for an innovative thorium-fueled approach are highly consequential for our future.

What if:

- We could have abundant drinking water for the 2+ billion people in water-stressed regions?
- We could grow crops in the desert, in the tundra, and on remote islands, and on the moon?
- We could turn seawater into net-carbon-neutral synthetic fuel for planes and ships and cars?
- We could reduce resource contention over fuel, food, and water, thereby increasing global stability?
- We could power and water our forward operating bases, reduce logistics casualties, create a high power battlegrid during campaigns, perform humanitarian aid missions, power medical support outposts, and leave behind microgrids & desalination for restoring the peace?

- We had a nuclear power that could consume the vaults of uranium nuclear waste, and leave behind rare nuclear byproducts that cure cancer and make radio telescopes?
- We could stack revenues such as electricity powering AI servers, desalination, synthetic fuels, and medical isotopes?
- We could power space habitats, labs, comms, transportation, robotics, probes, mining, and refueling operations?
- We could build the future we want, with plentiful freshwater, thriving agriculture, sustainable energy, and fuel for bold space exploration?

Throughout this book, we have explored a diverse range of innovations, from coastal cities striving for freshwater independence to high-tech greenhouses thriving in previously barren farmland. We have witnessed remote research reactors powering microgrids and moon-base prototypes testing life support systems fueled by molten salt. In every context, whether on Earth or beyond, the need for reliable, carbon-free energy, especially high-temperature thermal energy, has emerged as the driving force behind progress.

As the population grows, it's evident that without also increasing dependable energy, our systems for managing water, food, fuel, health, and security remain fragile. While thorium-powered molten salt reactors may not solve every problem, they represent a foundational technology. These reactors provide a compact, continuous, and carbonless source of heat capable of transforming scarcity into resilience and creating abundance.

This dependable energy can support freshwater production, promote sustainable agriculture, ensure consistent baseload power, facilitate the creation of synthetic fuels, and even enable oxygen liquefaction for space launch systems. By transforming power plants from single-output utilities into multi-product industrial hubs, we create a variety of engineering opportunities grounded in prototypes and decades of research.

We've observed how thorium energy can address critical gaps where existing infrastructures struggle. Coastal cities deal with saltwater intrusion into aquifers, while inland farms face reduced irrigation options. Data centers are proliferating faster than renewable sources like solar and wind can support. Additionally, defense planners and disaster-response teams are seeking energy solutions that are mobile, tamper-resistant, and independent of fragile fuel logistics.

Thorium-powered systems not only ensure energy security but also enhance water security, agricultural stability, energy-intensive manu-facturing, and life-support systems for off-world exploration. They unlock the value of waste heat, repurpose monazite mine tailings into clean fuel, and align energy generation with our environmental, economic, and planetary goals.

By combining advanced water filters with a steady MSR heat source, utilities can keep water prices stable and reduce dependence on diesel and unpredictable fuel markets. This innovation helps farmers maintain year-round crop production and utilize waste heat in green-houses, leading to more consistent crop yields and reduced carbon pollutions in the food chain.

Additionally, airlines and navies can use carbon-neutral synthetic fuels without relying on bio-crops that could instead produce food. A compact thorium reactor deployed on the Moon or Mars could support vital life systems and simplify the local production of fuel for transport and space missions.

Next Steps

Because the remaining challenges are more institutional than techni-cal, the next steps are evident:

1. Modernize Nuclear Safety and Siting Regulations

Today's rules are built around pressurized water reactors, which run under different heat and pressure settings compared to liquid-fueled molten salt system. To maintain safety and efficiency, the rules need to be adjusted to accommodate these differences. That way, we can support new ideas while keeping public trust.

2. Launch Demonstration Projects with Integrated Benefits

It's essential to develop pilot projects that offer more than basic power generation. These projects could include water purification plants for water-stressed regions, modular fertilizer facilities for local ammonia production, or combined electricity-and-isotope centers to support regional hospitals. or facilities that provide both electricity and medical isotopes to support regional hospitals. Such multi-purpose sites will not only showcase real-world performance but also offer valuable case studies for investors, insurers, and policy makers.

3. Expand the Talent Pipeline for Multidisciplinary Success

Molten salt reactors require expertise beyond nuclear engineering, including high-temperature metallurgy, corrosion chemistry, AI-assisted operations, and supply chain logistics. Universities, technical colleges, and national labs must adapt their curricula to prepare a new generation of reactor operators, maintainers, and inspectors for this advanced infrastructure.

4. Involve Communities as Co-Designers and Stakeholders

For infrastructure projects to succeed, communities must feel engaged rather than imposed upon. Developers should actively involve local citizens, landowners, utilities, and civic leaders from the outset. This includes transparent monitoring, revenue sharing where relevant, and responsiveness to local needs such as affordable water, job creation, and energy security.

5. Stack Multi-Stream Returns from a Single Infrastructure Investment: Why Smart Capital Is Moving Toward Multi-Revenue Infrastructure

Rather than measuring thorium projects solely by cash-on-cash return or payback period, investors should also consider their value in stabilizing water prices, reducing fertilizer imports, eliminating fuel convoys, and enabling the production of synthetic fuels. These systems lower carbon costs, strengthen local supply chains, and reduce strategic risks; all important factors to consider in any capital investment model.

Thorium-fueled MSRs can also generate real, recurring revenues across multiple high-demand markets. As outlined in earlier chapters, a single thorium reactor can simultaneously deliver electricity to AI data centers, provide heat and power for clean water production, produce synthetic fuel for transportation or export, and generate high-value medical isotopes. These aren't theoretical benefits. Each of these sectors represents a growing market, and thorium systems can serve them all from a compact, modular footprint.

These stacked revenue streams reduce exposure to commodity volatility, stabilize prices for essentials such as water and fertilizer, and create a diversified income model that is resilient across energy, agriculture, healthcare, and transportation. Beyond earnings, thorium projects deliver carbon reduction, domestic energy security, and geopolitical risk mitigation, intangible benefits that translate directly into long-term portfolio stability.

Thorium infrastructure isn't just cost-efficient. It's revenue-diversified, policy-aligned, and built to perform across multiple economic cycles. That's why smart capital is moving now. Not just to fund clean energy, but to own the future of multi-output infrastructure.

6. Establish a Framework for Energy-Space Collaboration

As countries and private companies grow their efforts in lunar missions and space infrastructure, energy systems need to advance as well. Compact, long-duration reactors that can support water collection, fuel creation, and basic life support will be vital for sustaining activities beyond Earth.

Collaboration among space agencies and energy companies is crucial for the design, development, and testing of space-based reactors, with protocols for transparency, emergency power-sharing, limits on territorial claims, and conflict resolution. This approach ensures equitable access to energy resources and promotes safe exploration.

How to Help

None of those tasks require you to be a reactor designer.

- A teacher can add the thorium fuel cycle and molten salt diagrams to the energy-transition module.
- A student could take a course on corrosion chemistry.
- A city planner can include nuclear RO desalination in the next water master plan.
- A farmer can ask suppliers about fertilizer whose hydrogen comes from emission-free heat.
- A philanthropist can fund a small modular power and water project in a remote area.
- A legislator can open a hearing on adding molten salt reactors to the clean energy portfolio and modernize the nuclear insurance statutes.
- An innovator could apply AI for improvements to nuclear chemistry.
- A supporter of thorium could share verified information to correct misconceptions online.

Each small intervention shortens the distance between the laboratory and the working infrastructure.

Building the Future We Want

This book advocates a future filled with plentiful freshwater, sustainable harvests, reliable energy, and opportunities for space exploration. While thorium-based systems do not guarantee this future, they make it technically and economically achievable within a generation if we decide to act. The tools are available, scientific research is published, and prototypes are in operation. Next comes a collective commitment, demonstrated through policy decisions, course enrollments, prototype funding, community meetings and posts that share the facts and encourage open conversation.

We can either continue to ration water, debate carbon allowances, patch up overworked power systems every summer, and delay the development of infrastructure needed for a planet with ten billion people. Or we can take a more deliberate approach by incorporating thorium into the broader clean-energy toolkit alongside wind, solar, batteries, and energy-saving measures. What happens next is a collective choice: a commitment to act faster, collaborate, and turn ideas into reality.

The science exists. The designs are published. The prototypes are built.

It's an engineering task, one of the most important of our time.

Let's get to work. Let's build the future worth having.

Additional Reading – and Watching

Want to dive deeper into thorium, molten salt reactors, and sustainable desalination? Here's a carefully curated mix of books, reports, studies, and video content to expand your understanding and keep the conversation going.

📚 Books and Reports

If you're new to nuclear science or want a clear, compelling case for thorium, start with **Robert Hargraves' Thorium: Energy Cheaper Than Coal**. It remains one of the most accessible and persuasive arguments for why thorium reactors are safer, cleaner, and more economically viable than traditional uranium-fueled designs.

Hargraves' more recent book, **New Nuclear is Hot!,** builds on that foundation with a timely analysis of how nuclear energy, including molten salt reactors, can help meet today's urgent climate and water security needs.

The **International Atomic Energy Agency (IAEA)** offers several cornerstone reports worth reading:

- Status of Molten Salt Reactor Technology
- The Thorium Fuel Cycle: Benefits and Challenges
- Near Term and Promising Long Term Options for the Deployment of Thorium-Based Nuclear Energy

These publications provide critical insights into the technical, regulatory, and policy considerations shaping thorium's global rollout. They also explore five distinct reactor configurations and assess their readiness and applicability in different parts of the world.

If water is your focus, check out the **Purdue University research led by David Warsinger** on batch reverse osmosis. His work outlines new designs that drastically reduce the energy cost of desalination, aligning perfectly with the promise of thorium reactors powering next-generation water purification systems.

🏛 Institutions and Advocacy Networks

Thorium Energy Alliance (TEA) continues to be a leading voice in thorium advocacy, providing education, organizing conferences, and lobbying for public and private investment in molten salt technologies.

Nuclear Innovation Alliance (NIA) focuses more broadly on advanced nuclear innovation, especially Small Modular Reactors (SMRs), and often includes key policy recommendations and industry case studies.

If you want the historical foundation, don't miss **Alvin Weinberg's autobiography, *The First Nuclear Era: The Life and Times of a Technological Fixer***. Weinberg pioneered molten salt reactor designs at Oak Ridge National Lab in the 1960s and was a visionary who saw nuclear not just as a power source, but as a solution to water scarcity, conflict, and long-term planetary stewardship.

💻 Recommended YouTube Channels and Videos

For clear and accessible explanations of thorium and molten salt technologies, check out these YouTube creators. They provide engaging visual content that makes learning easy.

Kirk Sorensen – Energy from Thorium

Start with Sorensen's iconic TED Talk: "Thorium: An Alternative Nuclear Fuel".

Follow up with in-depth technical walkthroughs on the **Energy from Thorium** YouTube channel, where he explains how molten salt reactors work, and why they matter.

Kyle Hill – "Why Thorium Could Be the Future of Nuclear Energy"

This well-reasoned and visually compelling breakdown delivers both scientific accuracy and narrative style. Hill explains the promise of thorium reactors while honestly exploring the challenges, perfect for first-timers and skeptics alike.

ColdFusion – "The Forgotten Fuel of the Future"

With millions of views, this documentary-style video offers a high-level introduction to thorium energy, its scientific roots, and its modern resurgence, told through sleek visuals and calm narration.

Science Time – "Molten Salt Reactors Explained"

Great for visual learners who want animated schematics and easy-to-grasp comparisons between thorium MSRs, traditional uranium reactors, and fusion alternatives.

The Thought Emporium – "Building a Molten Salt Reactor in My Garage"

For DIY tech fans or aspiring innovators, this independent experimenter shows what happens when thorium curiosity meets real-world tinkering.

Whether you're a policymaker, student, investor, or simply a curious mind, these resources offer something for every level of interest. This field is evolving quickly, with new research, prototypes, and policy shifts emerging every month. Staying informed and engaged is part of building the thorium-powered future we've envisioned together in these pages.

Let the journey continue on screen, on paper, and maybe even in your own lab.

Thank You for Listening

If you found this book thought-provoking, hopeful, or even just useful as a conversation-starter on clean energy and freshwater innovation, I'd be deeply grateful if you left a quick review.

Your honest feedback helps other readers discover this message and it signals to the world that there's growing interest in real, scalable solutions.

Even just a sentence or two makes a difference.

➡️ Please leave a review at the site where you attained this book.

For bonus content, updates, visuals, and clickable links to all the reports, studies, and resources cited in this book, visit our official companion site:

🌐 https://inov8r.com

We're building a community of innovators, engineers, policymakers, educators, and readers like you who see what's possible and want to help make it real.

Thank you for being part of that journey.

Bibliography

NOTE: A DOWNLOADABLE VERSION WITH HYPERLINKS IS AVAILABLE AT HTTPS://INOV8R.COM/

Alabama legislature, "advanced nuclear energy facilitation act," act no. 2025-112.

Alaska center for energy and power, "energy and food resilience in remote alaska," 2022 briefing paper.

Alvin weinberg, the first nuclear era: the life and times of a technological fixer (new york: aip press, 1994), 202.

Anduril autonomous systems: www.anduril.com

Australian academy of technology & engineering. Nuclear energy for australia. Melbourne, 2018.

Bhabha atomic research centre. "thorium fuel cycle in india: technology status and future roadmap." Barc annual report, 2023.

Boston dynamics robotics: www.bostondynamics.com

Breakthrough energy ventures. "climate-resilient infrastructure: investment themes." Investor memo, 2023.

Bwx technologies, "bwxt to build nuclear thermal propulsion reactor for darpa's draco," bwxt newsroom, july 26, 2023, https://www.bwxt.com/bwxt-to-provide-nuclear-reactor-engine-and-fuel-for-darpa-space-project/

California public utilities commission, "nuclear moratorium review brief20.

Canadian nuclear safety commission, 'heavy water and reactor design considerations', 2023.

Canadian nuclear safety commission, "vendor design review program overview," 2024.

Canadian nuclear safety commission. "heavy water and reactor design considerations." Ottawa, 2023.

Cern. "accelerator-driven systems for energy and waste transmutation." Workshop proceedings, 2022.

China national nuclear corporation, 'thorium msr pilot program status', 2025.

China national nuclear corporation. "thorium msr pilot program status." Beijing, 2025.

Chinese academy of sciences. "experimental thorium msr reaches initial operation in gansu." Cas bulletin, october 2023.

Cnsa mars reactor announcement: chinese academy of sciences bulletin, 2022

Coleridge, samuel taylor. The rime of the ancient mariner. 1834.

COLORADO GENERAL ASSEMBLY, "HOUSE BILL 1040: CLEAN ENERGY REFORM," 2025.

COMMONWEALTH FUSION SYSTEMS. "SPA`RC: A COMPACT, HIGH-FIELD, NET ENERGY FUSION DEVICE." ACCESSED MAY 2025. HTTPS://CFS.ENERGY/SPARC/

COPENHAGEN ATOMICS. "PARTNERSHIP WITH PAUL SCHERRER INSTITUTE FOR CRITICALITY DEMONSTRATION." PRESS RELEASE, NOVEMBER 20, 2023. HTTPS://WWW.COPENHAGENATOMICS.COM/NEWS

CYBERSECURITY AND INFRASTRUCTURE SECURITY AGENCY (CISA), "COLONIAL PIPELINE RANSOMWARE ATTACK," 2021 INCIDENT SUMMARY.

DARPA PROJECT PELE: https://www.darpa.mil/news-events/2022-09-22

DARPA, "DEMONSTRATION ROCKET FOR AGILE CISLUNAR OPERATIONS (DRACO)," DARPA NEWS, LAST MODIFIED JULY 26, 2023, https://www.darpa.mil/news-events/2023-07-26.

DEFENSE ADVANCED RESEARCH PROJECTS AGENCY (DARPA). "DECAY ON DEMAND." DARPA RESEARCH PROGRAMS, 2024. https://www.darpa.mil/program/decay-on-demand

DEFENSE ADVANCED RESEARCH PROJECTS AGENCY (DARPA). "SUNSPOT: SOURCES FOR ULTRAVIOLET NUCLEAR SPECTROSCOPY OF THORIUM." DARPA DEFENSE SCIENCES OFFICE, 2025. https://www.darpa.mil/research/programs/sunspot

DEPARTMENT OF ATOMIC ENERGY (INDIA), "KALPAKKAM HYBRID DESALINATION PROJECT," ANNUAL REPORT 2006–07 (MUMBAI: DAE, 2007), 112.

DEPARTMENT OF ATOMIC ENERGY (INDIA). "AHWR-LEU-300 STATUS UPDATE." GOVERNMENT OF INDIA, 2024.

EMIRATES NUCLEAR ENERGY CORPORATION. BARAKAH NUCLEAR ENERGY PLANT FACT SHEET. ABU DHABI, 2023.

EMIRATES NUCLEAR ENERGY CORPORATION. "UAE–FLIBE ENERGY THORIUM DESALINATION FEASIBILITY STUDY." PROJECT SUMMARY, 2024.

ENERGY RECOVERY INC. "OPTIMIZING REVERSE OSMOSIS PERFORMANCE THROUGH THERMAL INTEGRATION." TECHNICAL PAPER, 2022.

ENERGYFROMTHORIUM.COM, "LEARNING RESOURCES AND COMMUNITY DISCUSSIONS," 2025.

ESA THORIUM MSR FEASIBILITY STUDIES: EUROPEAN SPACE AGENCY PUBLICATIONS, 2021

EUROPEAN COMMISSION. STRATEGIC RESEARCH AGENDA FOR ADS. BRUSSELS, 2024.

EUROPEAN UNION, "HISTORY OF THE EUROPEAN COAL AND STEEL COMMUNITY," OFFICIAL RECORDS ARCHIVE, 2023.

EVOL/SAMOFAR. "EUROPEAN RESEARCH ON MOLTEN SALT REACTORS." EUROPEAN COMMISSION HORIZON 2020 REPORTS, 2022–2024.

FEDERAL ENERGY REGULATORY COMMISSION. ORDER DENYING REQUEST FOR INCREASED INTERCONNECTION CAPACITY AT SUSQUEHANNA NUCLEAR FACILITY. DOCKET NO. ER24-151, FEBRUARY 2024. HTTPS://WWW.FERC.GOV.

FLIBE ENERGY, "MSR FEASIBILITY AND SITING STUDY FOR AGRICULTURAL USE," INTERNAL WHITE PAPER, 2023.

Flibe energy. Press kit. 2024.

Flibe energy. "about us." Accessed 2025. Https://flibe.com

Food and agriculture organization of the united nations, *how to feed the world in 2050* (rome: fao, 2009).

Gates foundation. "water, sanitation & hygiene strategy overview." Seattle, 2024.

Generation iv international forum, 'molten salt reactor technology overview', updated 2023. Copenhagen atomics. "waste burner reactor." Copenhagen atomics, accessed june 2025. https://www.copenhagenatomics.com/technology/

Generation iv international forum, "global regulatory harmonization initiatives," 2023.

Generation iv international forum. "molten salt reactors: technology status paper." Paris, 2021.

German federal ministry for economic affairs and climate action, "energiewende: policy and progress," 2024.

Gleick, peter h. "water, drought, climate change, and conflict in syria." Weather, climate, and society 6 (2014): 331–340.

Government of iceland, national energy authority, "geothermal energy in iceland," 2023, .

Government of iceland, national energy authority, "geothermal energy in iceland," 2023, https://nea.is/geothermal.

Gulf chlor-alkali producers association. "market prices for solar-grade salt." Price bulletin, q4 2024.

Hargraves, robert. New nuclear is hot. Hanover, nh: createspace, 2025.

Hargraves, robert. Thorium: energy cheaper than coal. White river junction, vt: chelsea green publishing, 2012.

Iaea, "international research collaborations on advanced nuclear," 2024 progress summary.

Iaea, "smr regulators forum annual report," 2024

Iaea. Economic assessment of nuclear-powered desalination. Vienna, 2023.

Iaea. Introduction of nuclear desalination: a guidebook. Iaea-tecdoc-1444. Vienna, 2015.

Iaea. Management of uranium-233. Technical report series #481. Vienna, 2018.

Iaea. Near-term and promising long-term options for thorium deployment. Vienna, 2022.

Iaea. Status of molten salt reactor technology. Vienna, 2022.

Iaea. Status of nuclear desalination. Vienna, 2022.

Iaea. The thorium fuel cycle: benefits and challenges. Vienna, 2022.

Idaho state government, "inl partnerships and incentives," economic development summary, 2025.

Iea. Electricity 2024. Paris, 2024.

Iea. Energy security in a low-carbon world. Paris, 2021.

IEA. Net-zero roadmap. Paris: international energy agency, 2023.

Ignatiev, v., et al. "development of flibe for use in fusion-fission hybrids." Fusion engineering and design, vol. 136, 2018, pp. 1101–1107.

Indiana legislative services agency, "nuclear grid study committee report," 2025.

Institute of plasma physics, chinese academy of sciences. "fusion-fission hybrid reactor conceptual design report." Hefei, china, 2022.

International atomic energy agency, 'introduction to the use of thorium in nuclear reactors', iaea-tecdoc-1450 (2023).

International atomic energy agency, *introduction of nuclear desalination: a guidebook*, technical reports series no. 400 (vienna: iaea, 2000).

International atomic energy agency. Advances in small modular reactor technology developments. Vienna, 2022.

International desalination association. "global desalination market." Ida market brief, 2023.

International energy agency, "electricity 2024: analysis and forecast to 2026," iea report, january 2024.

International energy agency. Electricity 2024: analysis and forecast to 2026. Paris: iea, january 2024. Https://www.iea.org/reports/electricity-2024.

International membrane technology association. "guantanamo bay seawater ro desalination plant energy recovery systems." Case study, 2025.

Israel–jordan peace treaty. Annex ii, 1994.

Iter organization. "the iter project." Accessed may 2025. Https://www.iter.org/proj/inafewlines

J. D. Sethian et al., "the fusion-fission hybrid (ffh) reactor concept for advanced fuel cycles," los alamos national laboratory technical report la-ur-10-05645, 2010.

J.a. rakocy et al., "aquaponic production of tilapia and basil," *university of the virgin islands research bulletin*, 2004.

Jack spencer, nuclear revolution: powering the next generation (washington, dc: the heritage foundation press, 2024).

Jeff foust, "oklo seeks to expand microreactor applications to space and lunar surface," spacenews, april 19, 2023, https://spacenews.com/oklo-seeks-to-expand-microreactor-applications-to-space-and-lunar-surface/.

Jiangang li et al., "fusion–fission hybrid reactor (ffhr) research activities in china," nuclear fusion 59, no. 11 (2019): 112003. Https://doi.org/10.1088/1741-4326/ab3c61

K.a.care (king abdullah city for atomic and renewable energy). "small modular reactor road-map for desalination." Riyadh, 2024.

244

Kairos power. "kp-fhr project overview." Brief, 2024.

King abdullah city for atomic and renewable energy (k.a.care). "strategic energy transition report." Riyadh, 2023. Https://www.kacare.gov.sa/en/strategy/energy-transition-report

Kirk sorensen, remarks at thorium energy alliance conference 7 (teac7), 2016.ing," 2025.

Kirk sorensen, ted talk, 'thorium: an alternative nuclear fuel.'

Kirk sorensen. "fusion versus thorium." Presentation at teac10, thorium energy alliance, october 2023. Https://www.youtube.com/watch?v=video_id

Kirk sorensen. "high pain for low gain: why fusion isn't the answer." Energyfromthorium.com. Accessed may 2025. Https://energyfromthorium.com/fusion-critique/

Korean ministry of energy, "next-generation nuclear power development initiative annual report," 2024.

Kyle hill, "is thorium the future of nuclear power?", youtube video, 13:49, posted november 29, 2023, https://www.youtube.com/watch?v=7az1djhw7xw.

L.m. mortensen, "CO_2 enrichment in greenhouses," *agricultural and forest meteorology* 49, no. 3 (1990): 209–224.

Los alamos national laboratory msr salt chemistry research: https://www.lanl.gov

Los alamos national laboratory msr salt chemistry research: https://www.lanl.gov

Los alamos national laboratory. "fusion-fission hybrid systems for actinide management." La-ur-14-27624, 2014.

M.e. dry, "fischer–tropsch processes," *catalysis today* 71, no. 3–4 (2002): 227–241.

Massachusetts institute of technology. "nuclear science and engineering department: research highlights." Cambridge, 2025.

Mit nuclear energy futures lab. "thorium fuel cycle workshops." Cambridge, 2023.

Moisseytsev, a., and t. J. Heidet. "molten salt reactor design innovations enabled by ai." Argonne national laboratory, 2024. Https://www.anl.gov/article/molten-salt-design-with-ai

Moltex energy. "stable salt reactor – wasteburner." Accessed june 2025. Https://www.moltexenergy.com/technology/ssr-wasteburner

Montana legislature, "sb2024-14: modular nuclear support act," 2024.

Namibia water corporation ltd. And desert research foundation of namibia, "bethanie hybrid solar-powered desalination plant commissioned," project summary, july 2022. Https://www.drfnamibia.org.na/bethanie-desalination

Nasa artemis plan: 2020 plan for sustained lunar exploration and development

Nasa fission surface power system overview: https://www.nasa.gov/press-release/nasa-announces-fission-surface-power-project-awards

Nasa fission surface power system overview: https://www.nasa.gov/press-release/nasa-announces-fission-surface-power-project-awards

Nasa kilopower project: https://www.nasa.gov/directorates/spacetech/kilopower

Nasa kilopower project: https://www.nasa.gov/directorates/spacetech/kilopower

Nasa viper rover: https://www.nasa.gov/viper

Nasa viper rover: https://www.nasa.gov/viper

Natural resources canada, "energy support for remote communities," 2024 program brief.

Nature. "thorium reactors revisited." Commentary, 2012.

Navajo transitional energy company (ntec), "nuclear transition plan," 2023 announcement.

Norwegian ministry of education, "stem enrichment programs in upper secondary schools," 2020.

Nuclear energy agency, 'advanced nuclear fuel cycles and radioactive waste management', 2024.

Nuclear energy agency. Advanced nuclear fuel cycles and radioactive waste management. Paris, 2024.

Nuclear innovation alliance, "mission and impact," 2024 overview.

Nuclear innovation alliance. Advanced reactor licensing road-map. Washington, dc, 2022.

Nuclear innovation alliance. "enabling nuclear innovation: strategies for modernizing nuclear licensing." April 2023.

Nuscale power. "cost share termination notice." U.s. department of energy press release, 2023.

Oak ridge msr history: ornl technical memoranda, 1965–1975

Oak ridge national laboratory, 'molten salt reactor experiment final report', 1972.

Oak ridge national laboratory. Aircraft reactor experiment technical report. Oak ridge, tn, 1955.

Oak ridge national laboratory. Corrosion performance of hastelloy-n in molten fluoride salts. Materials report, 2020.

Oak ridge national laboratory. Global thorium supply memo. Internal memorandum, 2019.

Oak ridge national laboratory. Molten salt reactor experiment final report. Oak ridge, tn, 1972.

Oak ridge national laboratory. Msr fuel utilization briefing. Internal research memo, 2010.

Oak ridge national laboratory. Msre safety review. Oak ridge, tn, 1970.

Oecd nuclear energy agency, *nuclear education and training: from concern to capability* (paris: nea, 2021).

246

OHIO HOUSE OF REPRESENTATIVES, "ADVANCED REACTORS FOR RESILIENCE BILL," 2025.

PER PETERSON, JACOPO BUONGIORNO, AND ROBERT HARGRAVES. "WHY LICENSING FOR MOLTEN SALT REACTORS MUST CHANGE." NUCLEAR ENGINEERING REVIEW, FALL 2023.

PETER H. GLEICK AND YEMEN MINISTRY OF WATER AND ENVIRONMENT. NATIONAL WATER SECTOR STRATEGY. SANA'A, 2021.

QATAR ENVIRONMENT & ENERGY RESEARCH INSTITUTE. PILOT-SCALE MINERAL RECOVERY FROM DESALINATION BRINE. DOHA, 2024.

RUBBIA, CARLO, ET AL. "CONCEPTUAL DESIGN OF A FAST NEUTRON OPERATED THORIUM BURNER." CERN/AT/95-44, 1995.

SCK CEN. MYRRHA PROJECT SUMMARY. MOL, BELGIUM, 2024.

SORENSEN, KIRK. MEETING NOTES. AUTHOR'S PERSONAL FILES, 2013.

SORENSEN, KIRK. "THORIUM, AN ALTERNATIVE NUCLEAR FUEL." TED TALK. 2011. HTTPS://WWW.TED.COM/TALKS/KIRK_SORENSEN_THORIUM_AN_ALTERNATIVE_NUCLEAR_FUEL

SOUTH AFRICAN NUCLEAR ENERGY CORPORATION (NECSA), "ADVANCED REACTOR WORKSHOP SUMMARY," 2022.

SPACEX STARSHIP AVIONICS: STARBASE ENGINEERING UPDATE, MARCH 2024

SPENCER R. WEART. NUCLEAR FEAR: A HISTORY OF IMAGES. CAMBRIDGE: HARVARD UNIVERSITY PRESS, 1988.

STEENKAMPSKRAAL THORIUM LIMITED, "HTMR-100: HIGH TEMPERATURE MODULAR REACTOR OVERVIEW," COMPANY WHITE PAPER, ACCESSED JUNE 2025. HTTPS://WWW.THORIUM100.COM/

TASS RUSSIAN NEWS AGENCY, "ROSCOSMOS DEVELOPS SPACE TUG WITH NUCLEAR REACTOR," TASS, APRIL 13, 2021, HTTPS://TASS.COM/SCIENCE/1277477.

TENNESSEE DEPARTMENT OF ECONOMIC AND COMMUNITY DEVELOPMENT, "ADVANCED NUCLEAR COMMERCIALIZATION ZONE PROPOSAL," 2025.

TERRAPOWER. "OUR MISSION." TERRAPOWER.COM, ACCESSED JUNE 2025. HTTPS://WWW.TERRAPOWER.COM/ABOUT

TERRAPOWER. "TERRAPOWER STATEMENT ON THE NATRIUM REACTOR PROJECT." TERRAPOWER NEWSROOM, NOVEMBER 2023.

TERRELL, JEFF. "COPENHAGEN ATOMICS: THE DANISH STARTUP BUILDING MASS-PRODUCED THORIUM REACTORS." MEDIUM, MARCH 5, 2023. HTTPS://MEDIUM.COM/@TERRELLJEFF/COPENHAGEN-ATOMICS-THORIUM-VISION

TESLA OPTIMUS ROBOT ANNOUNCEMENT: HTTPS://EN.WIKIPEDIA.ORG/WIKI/OPTI MUS_(ROBOT)

TESLA OPTIMUS ROBOT ANNOUNCEMENT: HTTPS://EN.WIKIPEDIA.ORG/WIKI/OPTI MUS_(ROBOT)

TEXAS ENERGY COMMISSION, "PUBLIC COMMENT REQUEST: CLEAN HYDROGEN FRAMEWORK," 2025.

THAILAND INSTITUTE OF NUCLEAR TECHNOLOGY (TINT), "THORIUM RESEARCH COLLABORATION," POLICY BRIEF, 2024.

THERMAL MANAGEMENT: INCORPORATE RADIATOR ARRAYS USING SODIUM HEAT

PIPES OR LIQUID METAL LOOPS, AS EFFICIENT HEAT REJECTION OCCURS
SOLELY THROUGH RADIATION.

THORCON INTERNATIONAL, "THORCON AND PLN SIGN FRAMEWORK AGREEMENT
FOR 500 MW TMSR-500," NEWS RELEASE, FEBRUARY 6, 2024

THORIUM CANADA, "COMMUNITY ENGAGEMENT FOR MICROGRID SITING: ALBERTA
INDIGENOUS PILOT," 2023 FIELD REPORT.

THORIUM ENERGY ALLIANCE, "ABOUT US," HTTPS://THORIUMENERGYAL
LIANCE.COM.

THORIUM ENERGY ALLIANCE, "ABOUT US," HTTPS://THORIUMENERGYAL
LIANCE.COM.

THORIUM ENERGY ALLIANCE. "POLICY BRIEF: THORIUM FOR SECURE ENERGY AND
WATER." 2024.

THORIUM ENERGY WORLD. "INTERVIEW WITH THOMAS JAM PEDERSEN OF COPEN-
HAGEN ATOMICS." THORIUM ENERGY ALLIANCE, SEPTEMBER 12, 2022.

U.S. ARMY LOGISTICS UNIVERSITY. "FUEL CONVOY CASUALTY STATISTICS,
OEF/OIF." FORT LEE, VA, 2014.

U.S. DEPARTMENT OF DEFENSE. "PROJECT PELE: MOBILE MICROREACTOR DEVEL-
OPMENT UPDATE." 2025.

U.S. DEPARTMENT OF ENERGY. COASTAL RESILIENCE OPTIONS REPORT. WASHING-
TON, DC, 2024.

U.S. DEPARTMENT OF ENERGY. MOLTEN SALT REACTORS FACT SHEET. OFFICE OF
NUCLEAR ENERGY, 2024.

U.S. GEOLOGICAL SURVEY. COLORADO RIVER BASIN WATER SUPPLY AND DEMAND
STUDY: 2022 UPDATE TO CONGRESS. CIRCULAR 1491.

UK OFFICE FOR NUCLEAR REGULATION, "TECHNOLOGY-NEUTRAL ADVANCED
REACTOR LICENSING," REGULATORY UPDATE, 2023.

UN WATER. WORLD WATER DEVELOPMENT REPORT. PARIS: UNESCO, 2024.

UNITED NATIONS DEPARTMENT OF ECONOMIC AND SOCIAL AFFAIRS, *WORLD
POPULATION PROSPECTS 2022* (NEW YORK: UNITED NATIONS, 2022).

UNITED NATIONS GENERAL ASSEMBLY. TRANSFORMING OUR WORLD: THE 2030
AGENDA FOR SUSTAINABLE DEVELOPMENT. A/RES/70/1, 2015.

US DEPARTMENT OF DEFENSE, 'PROJECT PELE: MOBILE MICROREACTOR DEVELOP-
MENT UPDATE', 2025.

US DEPARTMENT OF ENERGY, 'MOLTEN SALT REACTORS', OFFICE OF NUCLEAR
ENERGY FACT SHEET, 2024.

US DEPARTMENT OF ENERGY, ADVANCED MANUFACTURING OFFICE, *INDUSTRIAL
DECARBONIZATION ROADMAP* (WASHINGTON, DC: DOE, 2021).

US DEPARTMENT OF ENERGY, IDAHO NATIONAL LABORATORY, "HIGH TEMPERA-
TURE ELECTROLYSIS FOR HYDROGEN PRODUCTION," ACCESSED 2023.

US DEPARTMENT OF ENERGY, OFFICE OF NUCLEAR ENERGY, "ALASKA MICRORE-
ACTOR INITIATIVE," PRESS RELEASE, APRIL 2022.

US DEPARTMENT OF ENERGY, OFFICE OF NUCLEAR ENERGY, "PROJECT PELE:
MOBILE MICROREACTOR," HTTPS://WWW.ENERGY.GOV/NE/ARTICLES/PROJECT-
PELE-MOBILE-MICROREACTOR.

US DEPARTMENT OF ENERGY, "HALEU AVAILABILITY AND COMMERCIALIZATION STRATEGY REPORT," MARCH 2025.

US NUCLEAR REGULATORY COMMISSION. ADVANCED REACTOR CONTENT OF APPLICATION PROJECT (ARCAP), OFFICE OF NEW REACTORS, 2024. HTTPS://WWW.NRC.GOV/REACTORS/NEW-REACTORS/ADVANCED/ARCAP.HTML

US NUCLEAR REGULATORY COMMISSION. CYBERSECURITY REGULATORY GUIDE FOR DIGITAL INSTRUMENTATION AND CONTROL SYSTEMS, REV. 1, JANUARY 2024.

US SENATE, "SENATE RESOLUTION 155: NUCLEAR DESALINATION STUDY," CONGRESSIONAL RECORD, 1967.

USNC-TECH, "USNC-TECH AND NASA COLLABORATE ON SPACE NUCLEAR POWER SYSTEMS," ULTRA SAFE NUCLEAR NEWSROOM, DECEMBER 14, 2022, HTTPS://WWW.USNC.COM/NEWS/USNC-TECH-AND-NASA-COLLABORATE-ON-SPACE-NUCLEAR-POWER-SYSTEMS.

WEINBERG, ALVIN M. THE FIRST NUCLEAR ERA: THE LIFE AND TIMES OF A TECHNOLOGICAL FIXER. NEW YORK: AMERICAN INSTITUTE OF PHYSICS, 1994.

WHITE HOUSE, EXECUTIVE ORDERS 14089, 14090, AND 14091, MAY 23, 2025, HTTPS://WWW.WHITEHOUSE.GOV.

WHITE HOUSE, EXECUTIVE ORDERS 14089, 14090, AND 14091, MAY 23, 2025, HTTPS://WWW.WHITEHOUSE.GOV.

WORLD BANK. RED SEA–DEAD SEA WATER CONVEYANCE STUDY PROGRAM: ENVIRONMENTAL AND SOCIAL ASSESSMENT – EXECUTIVE SUMMARY. WASHINGTON, DC, 2014.

WORLD INSTITUTE FOR NUCLEAR SECURITY. SAFEGUARDS FOR LIQUID-FUEL REACTORS. VIENNA, 2023.

WORLD METEOROLOGICAL ORGANIZATION. STATE OF GLOBAL WATER RESOURCES 2023. GENEVA, 2024.

WORLD NUCLEAR ASSOCIATION, 'NUCLEAR FUEL CYCLE', ACCESSED 2025.

WORLD NUCLEAR ASSOCIATION, "FRANCE: PUBLIC SUPPORT FOR NUCLEAR ENERGY," 2023.

Glossary of Acronyms and Abbreviations

ADS stands for Accelerator-Driven System. CERN and the Belgian research program MYRRHA use this term for sub-critical reactors powered by a proton accelerator, an approach often paired with thorium to curb long-lived waste.

AEC remains the familiar Atomic Energy Commission, the historical US agency that launched many early reactor programs.

AECL is Atomic Energy of Canada Limited, the crown corporation that pioneered the CANDU line.

AI refers to Artificial Intelligence, a thread that runs through Chapters 4 and 10 in connection with reactor monitoring and plasma control.

APR designates the Korean Advanced Power Reactor, cited in the Barakah case study.

BBC is the British Broadcasting Corporation, cited in Chapter 5 on lunar reactor concepts.

CANDU means CANada Deuterium Uranium, the heavy-water reactor fleet exported worldwide.

CERN is the European Organization for Nuclear Research, noted for its ADS work.

CNL identifies Canadian Nuclear Laboratories, now steward of Chalk River.

CNSC is the Canadian Nuclear Safety Commission, Canada's independent nuclear regulator.

CNSA abbreviates the China National Space Administration, mentioned in the space-power discussion.

DOE is the United States Department of Energy.

ENEC refers to the Emirates Nuclear Energy Corporation, owner of the Barakah plant.

ESA stands for the European Space Agency.

FANR is the Federal Authority for Nuclear Regulation of the United Arab Emirates.

GIF denotes the Generation IV International Forum.

GTMO is shorthand for the US naval base at Guantánamo Bay.

HEU means Highly Enriched Uranium.

IAEA is the International Atomic Energy Agency.

IDA represents the International Desalination Association.

IEA refers to the International Energy Agency.

ISRU stands for In-Situ Resource Utilization, the practice of drawing fuel and consumables from extraterrestrial soils.

ITER is the International Thermonuclear Experimental Reactor.

KEPCO is the Korea Electric Power Corporation, parent of the APR-1400 design.

LBE means Lead–Bismuth Eutectic, a coolant alloy discussed alongside molten salts.

LCF in this manuscript is Lattice Confinement Fusion, NASA's exploratory fast-fission concept, not Life-Cycle Funding.

LCOE still indicates the Levelized Cost of Electricity.

LFTR remains the Liquid Fluoride Thorium Reactor.

LNG is Liquefied Natural Gas, cited in the GTMO retrofit.

MED refers to Multiple-Effect Distillation, a thermal desalination method.

MIT stands for the Massachusetts Institute of Technology.

MOX is Mixed-Oxide fuel.

MSF means Multi-Stage Flash distillation.

MSR is Molten Salt Reactor, while MSRE is the historic Molten Salt Reactor Experiment at Oak Ridge.

MYRRHA is the Multi-purpose hYbrid Research Reactor for High-tech Applications, Belgium's pilot ADS facility.

NASA is the National Aeronautics and Space Administration.

NIA is the Nuclear Innovation Alliance.

NRCan shortens Natural Resources Canada, referenced in Chapter 9.

ONR is the United Kingdom's Office for Nuclear Regulation.

ORNL stands for Oak Ridge National Laboratory, source of many molten-salt milestones.

PBR designates the Pebble Bed Reactor.

PDF is Portable Document Format.

RBMK denotes the Soviet-era channel-type reactor.

RO stands for Reverse Osmosis desalination.

ROI means Return on Investment.

SCK CEN shortens the Belgian Studiecentrum voor Kernenergie, partner in MYRRHA.

SDGs are the United Nations Sustainable Development Goals, invoked in Chapter 7's water-security section.

SMR is Small Modular Reactor.

SPARC, the compact high-field tokamak under construction in the United States, expands to Soonest / Smallest / Private-funded / Affordable / Robust / Compact.

STI-DOC is the Scientific and Technical Information Document series.

TEA marks the Thorium Energy Alliance.

TED refers to the Technology Entertainment Design conference series that hosted Kirk Sorensen's talk.

UAE is the United Arab Emirates.

UK stands for the United Kingdom, while UN is the United Nations and US remains the United States.

VS is used only as the comparative "versus."

WMO is the World Meteorological Organization, cited in the State of Global Water Resources report.

WNA means World Nuclear Association, the industry body whose research appears in several chapters.

ZLD concludes the list with Zero-Liquid-Discharge brine management.

Appendix: A Technology Companion for Chapter 10: Powering the Final Frontier: Thorium MSRs and the Architecture of Space Civilization

The Challenge

Space is both inviting and unwelcoming. A blank canvas of stars, silence, and starkness. In space, energy is survival. The farther we go, from the lunar poles to the Martian plains, from asteroids to exoplanets, the more critical energy becomes. Every habitat, rover, lab, drill, and printer rely on electricity. Unlike Earth, there are no backup power grids or fossil fuels available. While solar panels are effective in orbit, they fail in Martian dust storms. Batteries have limited capacity, and RTGs offer only a trickle of power, which is inadequate for sustaining a settlement.

"Now, I spent 10 years working at NASA... and we had to think about how we would provide energy for this very unique community. There's no coal on the Moon. No petroleum. No atmosphere. Solar power fails for two weeks during the lunar night. Nuclear energy was really the only choice."

— Kirk Sorensen TED Talk, Thorium, an
alternative nuclear fuel[1]

A scalable, continuous power source is imperative in space. Thorium molten salt reactors (MSRs) can fulfill this role. To venture outward means building from scratch in one of the most hostile environments imaginable. Yet within this vacuum lies humanity's greatest opportunity not just to explore, but to establish, endure, and thrive as a space-based community.

In 2025, the US Presidential Executive Order 13972 promoted the development and deployment of small modular reactors (SMRs) for space exploration.

But what would deploying reactors in space involve? Citing the known facts and applying some speculative imagination, let's envision a future that uses thorium SMRs for various missions, taking bold steps from launch operations to establishing lunar colonies, facilitating space travel to Mars, conducting mining, manufacturing for self-sufficiency, launching deep space probes and exploring beyond.

Space Thorium MSR Roadmap: Step by Step Development Framework

THORIUM APPLICATIONS ROADMAP:

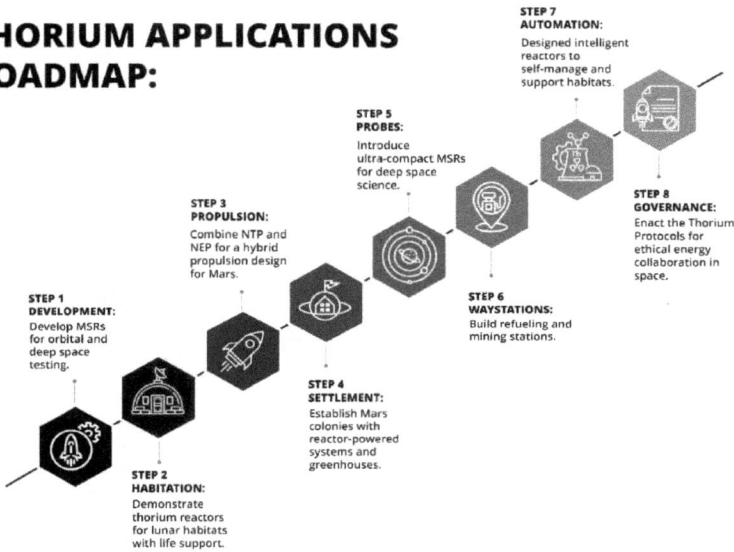

STEP 7
AUTOMATION:
Designed intelligent reactors to self-manage and support habitats.

STEP 5
PROBES:
Introduce ultra-compact MSRs for deep space science.

STEP 3
PROPULSION:
Combine NTP and NEP for a hybrid propulsion design for Mars.

STEP 8
GOVERNANCE:
Enact the Thorium Protocols for ethical energy collaboration in space.

STEP 1
DEVELOPMENT:
Develop MSRs for orbital and deep space testing.

STEP 6
WAYSTATIONS:
Build refueling and mining stations.

STEP 4
SETTLEMENT:
Establish Mars colonies with reactor-powered systems and greenhouses.

STEP 2
HABITATION:
Demonstrate thorium reactors for lunar habitats with life support.

Step 1: Preparing for Launch: Engineering for the Void

The Vision

Space is an unforgiving realm. No air. No pressure. No gravity worth counting on. And most challenging for a power system: no atmosphere to convectively cool waste heat. These conditions necessitate a radical redesign of terrestrial energy systems before nuclear technologies, including MSRs, can safely and effectively operate beyond Earth.

Unlike pressurized water reactors (PWRs), which rely on high-pressure steam and heavy cooling systems, MSRs operate at atmospheric pressure and use a liquid fluoride salt mixture as both fuel carrier and coolant. This makes them an excellent candidate for space, provided several adaptations are made to account for the vacuum, radiation, and thermal dynamics of space.

The "Scalable Space MSR"

There are core technology enablers for optimizing thorium MSRs for space applications. To reduce repetition, we will refer to this as the "Scalable Space MSR," which supports modular scaling from 10 kW probe systems to multi-hundred-kW station hubs.

Here is the Scalable Space MSR reference architecture:

FUEL COMPOSITION: The reactor uses Thorium-232 with a U-233 starter load. It can start using a neutron source or cycles that are replenished with resources from in-situ utilization (ISRU).

COOLANT: The coolant is FLiBe (a mixture of lithium fluoride and beryllium fluoride), which can operate steadily at about 700°C.

CONTAINMENT MATERIALS: We use Hastelloy-N or similar corrosion-resistant alloys. The system is designed to be unpressurized.

MODERATOR: We use either a graphite array or a salt-embedded moderation system, depending on the reactor's size.

POWER CONVERSION: The reactor converts power using a Brayton cycle, which is preferred, or a Stirling engine in a closed-loop system.

THERMAL MANAGEMENT: It features high-emissivity radiators with sodium heat pipes or liquid metal loops to manage heat.

PASSIVE SAFETY: The reactor features a drain system with

A FREEZE PLUG, DESIGNED TO OPERATE SAFELY IN ZERO OR LOW GRAVITY ENVIRONMENTS.

AUTONOMOUS CONTROL: AN AI SYSTEM MANAGES THE OPERATION OF THE REACTOR AND PROVIDES PREDICTIVE DIAGNOSTICS.

REDUNDANCY: THERE ARE DUAL COOLANT PUMPS AND DRAIN RESERVOIRS THAT CAN BE SWITCHED OUT WHILE OPERATING.

SHIELDING: THE REACTOR USES MODULAR SHIELDING THAT CAN INCLUDE WATER TANKS OR BARRIERS MADE FROM REGOLITH (THE DUST AND BROKEN ROCK ON THE MOON OR MARS).

The graphite moderator helps to use neutrons more efficiently, allowing more efficient use of fuel and reducing the amount of fissile material required. FLiBe coolant has good thermal stability and low vapor pressure, which helps it transfer heat effectively in a vacuum. Together, these choices support compact, modular, and autonomous operation, ideal for space applications.

Scalable Space MSR Architecture

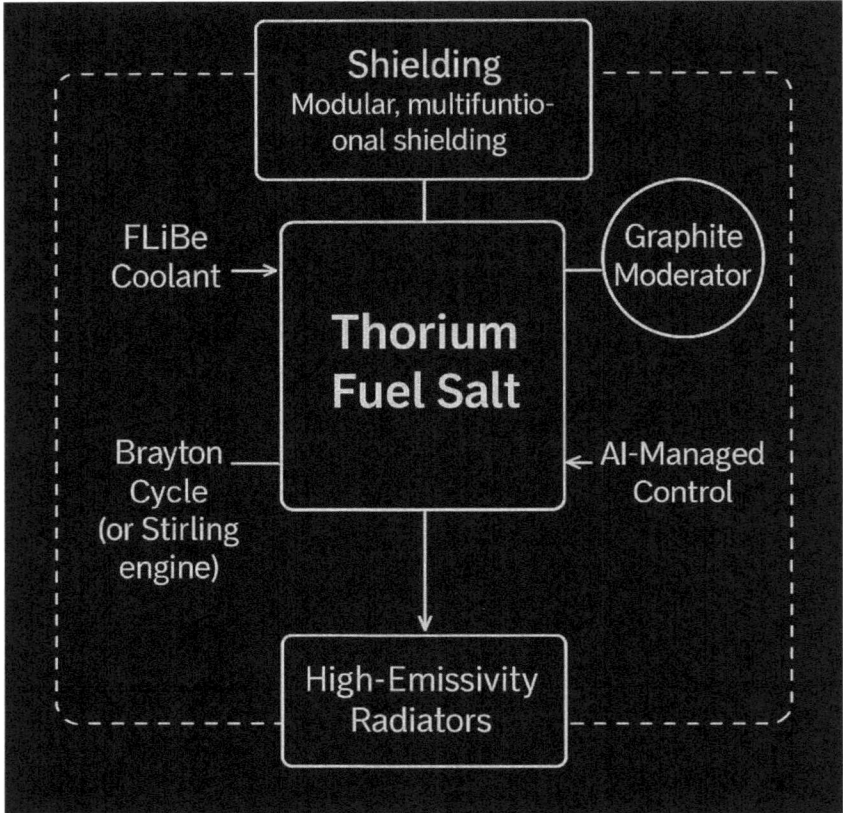

The "Space Tech Toolkit": Starting Points of Technology Applicable to Space

To support a space architecture powered by thorium and designed for multiple environments, here is a "Space Tech Toolkit" that categorizes key space technologies by their functions.

This toolkit of enabling technologies will be helpful for the rest of this chapter.

1. Reactor and Fuel Systems

NASA KILOPOWER[2]: DEMONSTRATED SMALL-SCALE, AUTONOMOUS URANIUM FISSION REACTORS (1–10 kWE) WITH STIRLING ENGINES AND PASSIVE STARTUP.

COPENHAGEN ATOMIC'S MOLTEN SALT REACTORS[3] ARE COMPACT ENOUGH TO FIT INSIDE A STANDARD 40-FOOT SHIPPING CONTAINER, BUT POWERFUL ENOUGH TO PRODUCE AROUND 100 MEGAWATTS OF THERMAL ENERGY AT HIGH TEMPERATURES AND LOW PRESSURE.

NASA FISSION SURFACE POWER (FSP)[4]: PRODUCING SCALABLE 40 kWE SYSTEMS WITH COMMERCIAL PARTNERS; ADAPTABLE TO FUTURE THORIUM AND MOLTEN SALT DESIGNS.

CNSA (CHINA)[5]: EARLY-STAGE DEVELOPMENT OF 1 MWE CLASS REACTORS WITH STATED THORIUM MSR AMBITIONS FOR PLANETARY BASES.

ESA MOON/MARS MSR STUDIES[6]: EMPHASIZING SALT CHEMISTRY, HIGH-TEMPERATURE RADIATOR INTEGRATION, AND TRANSPORTABILITY.

2. Heat Rejection and Thermal Management

PROJECT PELE AND KILOPOWER: BOTH INCORPORATE PASSIVE THERMAL TRANSPORT AND RADIATOR LOOP TESTING.

ESA FEASIBILITY STUDIES: FOCUSED ON LOW-GRAVITY RADIATOR EFFICIENCY AND MODULAR HEAT LOOP DESIGN.

3. AI and Autonomous Control Systems

KILOPOWER PROTOTYPE: DEMONSTRATED AUTONOMOUS STARTUP AND SHUTDOWN SEQUENCES.

SpaceX Starship Avionics: Advanced real-time fault detection and machine-learning-based control algorithms.

NASA VIPER & DART Missions: Onboard navigation and obstacle avoidance via AI guidance.

4. Fuel Handling and Salt Chemistry

Los Alamos National Laboratory[7]: MSR corrosion studies and salt purification research inform ISRU-compatible designs.

Oak Ridge National Laboratory (historical)[8]: Original source of MSR fuel loop and reprocessing experimentation.

5. Robotics and Mobility Platforms

Anduril Industries[9]: A leading developer of autonomous defense and mobility systems, including AI-powered drones and sensor platforms with potential for planetary adaptation.

Palladyne AI (formerly Sarcos Robotics): Now focused on AI software to support intelligent robotic behaviors and perception in mobile platforms.

Boston Dynamics[10]: Produces adaptable terrestrial robots capable of dynamic movement and terrain navigation.

Tesla Optimus[11]: In March 2025, Elon Musk announced a plan to send a Tesla Optimus humanoid robot to Mars onboard a SpaceX Starship in 2026. Optimus is designed for multi-role automation and mobility in harsh environments.

VIPER Rovers[12] **& NASA Artemis**[13] : Lunar rovers designed for drilling, regolith handling, and volatile resource extraction.

6. In-Situ Manufacturing and Additive Fabrication

Relativity Space: Demonstrating launch-grade 3D printing of rocket structures; scalable to field-fabricated infrastructure.

NASA ISRU Programs: Including regolith-based printing, sintering, and closed-loop parts manufacturing.

7. Power Conversion Systems

Kilopower: Uses Stirling engines with proven conversion efficiencies.

Project Pele: Integrating closed Brayton cycle turbines for modular power scaling.

8. Launch and Payload Integration

SpaceX & Blue Origin: Developing large-volume payload platforms and robotic deployment services for nuclear payloads.

Intuitive Machines: Specializing in lunar and planetary cargo delivery systems.

This toolkit provides a structured reference for each step in the thorium MSR development roadmap. Technologies cited are real, demonstrated, or in active prototype development, and provide a tangible launchpad for space-rated reactor infrastructure.

Space Tech Toolkit Capability Readiness Matrix

Capability	Readiness Level
Reactor and fuel systems	Prototype stage
Heat rejection and thermal management	Prototype stage
AI and autonomous cotrols	Operational
Fuel handling and salt chemistry	Early pathinding
Power conversion systems	Prototype stage
Launch and payload integration	Speculative but plausible

The Mission

Demonstrate MSR operations in Earth orbit and deep space analogs. Develop power generation systems and life support modules that rely entirely on MSRs in zero-gravity, zero-atmosphere conditions.

Power Load Requirements

Scalable Space MSR: Scale 10-40 kWe demonstration units for orbital and cislunar testbeds.

Relevant Space Tech Toolkit Capability

Reactor and Fuel Systems: NASA's Kilopower is in operation, while Copenhagen Atomic's reactors and NASA's FSP are still in the prototype stage.

Heat Rejection and Thermal Management: Project Pele's radiator testing reached prototype stage, and ESA is studying radiator concepts.

AI and Autonomous Controls: Kilopower's startup protocols and SpaceX's avionics are both operational.

Power Conversion Systems: The Kilopower Stirling engines are operational, while Pele's Brayton cycle turbines are in the prototype stage.

Stirling engine for NASA Kilopower

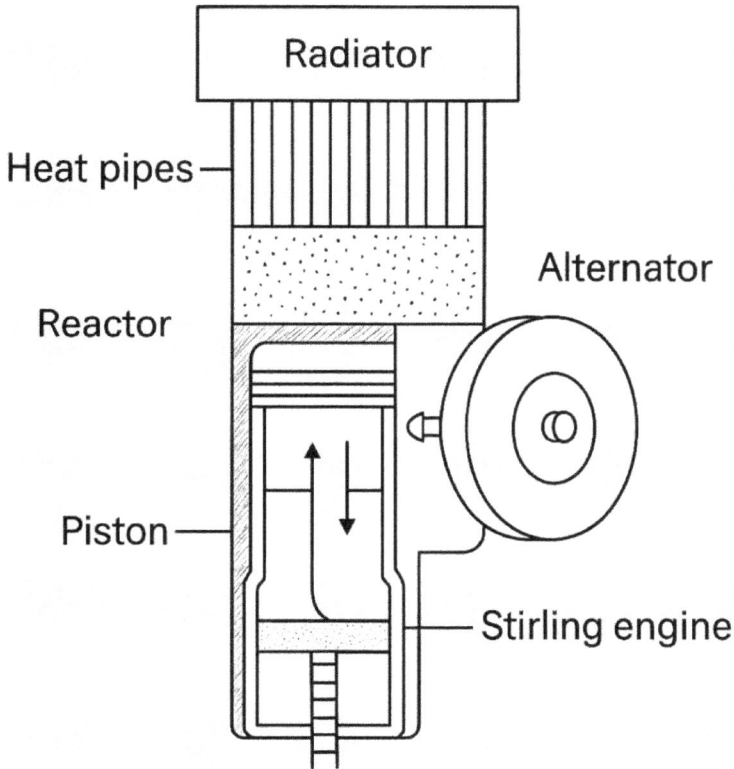

Launch & Integration: SpaceX and Blue Origin can handle payloads effectively, as they are both operational.

Innovation Needed

In a vacuum, heat can only be removed through radiation, which is heavily dependent on temperature. Modular MSRs, with outlet temperatures around 700°C, have a significant advantage over lower-temperature systems, such as pressurized water reactors (PWRs) or radioisotope thermoelectric generators (RTGs). Their higher operating temperature allows for more efficient radiative cooling. However, effective heat rejection still requires optimized radiator systems. Radiator fins or looped sodium heat pipes will need to be designed for weight, surface area, and dust resilience, especially on the Moon, where sharp regolith particles can degrade performance over time.

Let's compare the available power source options:

Space Power Sources: Solar vs RTG vs. Thorium MSR

Power Source	Efficiency	Output	Mass/Volume	Maintenance	Notes
Solar	Variable (5–25%)	Limited by insolation	Large surface area	Needs dust removal, battery storage	Ineffective during lunar night or Mars dust storms
RTGs	~6%	Low (100–500 W)	Compact	Fixed output, no moving parts	Best for deep space; insufficient for habitats
MSRs (Thorium)	~40–45%	Scalable (kW to MW)	Compact/modular	Passive safety, autonomous	Operates during lunar night; adaptable to all environments

A Reactor for the Final Frontier: Designing a Space-Rated MSR

Picture this: a molten salt reactor, humming quietly on the surface of the Moon. Or orbiting silently above Mars, powering life support systems and scientific instruments in the vacuum of space. But here's the catch: we can't just take a reactor designed for Earth and launch it into orbit. Space is a hostile, unforgiving place. No atmosphere. No gravity. No second chances. That's why space-rated molten salt reactors need a fundamental redesign from their Earth-bound cousins.

There are six key adaptations needed to make a molten salt reactor function in space.

First, there's the challenge of containment. On Earth, reactors are built to handle pressure. In space, we flip the script. These reactors must operate at atmospheric pressure, using an advanced alloy called Hastelloy-N. Why? Because it resists corrosion from the reactor's molten fluoride salts, even after years of high heat and radiation.

Second, we need fuel flexibility. A space MSR will use thorium-232 as the primary fuel, but to get things started, it carries a small charge of uranium-233. To ensure reliable startup in microgravity, optional neutron sources may be added as well.

Third, there's thermal management. On Earth, we get rid of waste heat through air and water. But in space? There's no air. So we rely entirely on radiation: literally radiating heat away into the black void. This means building massive radiator arrays using heat pipes filled with sodium or other liquid metals. These radiators have to be efficient, dust-resistant, and built to last.

Fourth, coolant circulation. Gravity-fed systems won't work in zero-g. So, we turn to mechanical or electromagnetic pumps to keep the coolant flowing and the core stable.

Fifth, we need to design for passive safety, even in microgravity. On Earth, molten salt reactors rely on a freeze plug, essentially a solid stopper that melts if the reactor overheats, allowing the fuel to safely drain. In space, that mechanism must work without gravity. That means engineering a zero-g-compatible freeze plug that functions reliably in all orientations.

And finally, we come to autonomy. There are no maintenance teams floating outside Jupiter. These reactors must run themselves. That means AI-driven control systems, predictive maintenance software, and remote diagnostics, all hardened against space radiation.

Without these six adaptations, even the best-designed MSR would be doomed in space. Our Scalable Space SMR design addresses every one of these challenges, including those in the reactor's secondary systems. Because adapting the core is only part of the job. The rest of the system needs a makeover, too.

Take heat rejection, for instance. On Earth, we rely on convection and conduction. In space, it's radiation or nothing. That's why our radiator panels are enormous, highly emissive, and designed to repel dust and micrometeoroids.

Power conversion? Forget steam turbines. In space, we favor Brayton or Stirling cycle engines: sealed, high-efficiency systems that work well in microgravity and don't rely on boiling water.

Materials and handling? There's no oxygen in space, so components must resist vacuum outgassing, that is when materials slowly shed molecules in a vacuum. Everything must stay structurally sound across extreme hot and cold cycles.

And then there's radiation shielding. It can't be static. It must be modular, maybe even dual-purpose. For example, we can design water tanks that provide shielding for the crew while also supplying the life support system.

All of this is more than just engineering. It's reimagining how we power civilization beyond Earth, one reactor at a time.

Adapting the Reactor's Unsung Heroes: The Secondary Systems

Designing the reactor core is only half the story. What often gets overlooked, but is just as critical, are the secondary systems. These are the support components that make the entire system work: the radiator panels, power converters, materials, and shielding. On Earth, they're built to work in a gravity field, surrounded by atmosphere. In space? Everything changes.

We will start with heat rejection, the most important and the most difficult challenge. On Earth, we get rid of heat through air and water, through convection. But in the vacuum of space, there is no air to carry heat away. There's only radiation. That means we need enormous radiator panels to shed the reactor's thermal energy. These panels must be lightweight but durable, highly emissive, and dust-resistant, especially if they're destined for the lunar surface or Mars, where fine regolith clings to everything. And because there's no wind or weather, their cooling performance must be entirely passive: no airflow, no fans, just pure infrared glow into the blackness.

Next, we come to power conversion. On Earth, we often use steam turbines in a Rankine cycle, turning boiling water into power. That won't work in space. Instead, we favor Brayton or Stirling cycles: closed-loop systems that run without water, without atmospheric pressure, and without gravity. They use compressible working fluids like helium or supercritical CO_2, cycling through expansion and compression chambers to spin a generator. These systems are sealed, silent, and capable of operating in low-gravity environments for years at a time.

Then there's the matter of materials and handling. On Earth, we take atmospheric oxygen for granted - it keeps most materials stable. In space, there is none. Materials outgas in vacuum, shedding molecules that can foul sensors or coat optics. Thermal

271

expansion and contraction are more extreme, especially on the Moon, where surface temperatures can swing hundreds of degrees. So every gasket, pipe, and panel must resist corrosion, outgassing, and structural fatigue. Materials must be space-rated: designed to survive a wide thermal envelope, intense radiation, and years of wear in silence.

And finally, there's radiation shielding. In space, radiation doesn't come from just one direction. It comes from the Sun. From cosmic rays. From the reactor itself. So, shielding must be modular, directional, and multifunctional. What's a smart solution we can borrow from submarines? Water tanks that surround the reactor serve a dual purpose, acting as both life support and protection. Water is a phenomenal radiation absorber. So, while the crew uses it to drink and recycle, it can also block neutrons and gamma rays from the reactor. This type of dual-use engineering is essential in space, where every gram counts and every system must serve multiple purposes.

These adaptations are vital. Without them, even the best-designed reactor won't survive the rigors of launch, orbit, or off-world operation. But with them, we enable crew to live, work, and thrive beyond Earth.

Towards a Space Energy Architecture

Molten salt reactors are crucial for creating self-sustaining outposts and enabling advanced activities such as in-situ resource utilization (ISRU), manufacturing in space, and deep-space missions. To move forward, we need to establish testing platforms for MSRs in orbit, integrate these habitats autonomously, and develop systems that share fuel and heat effectively with other technologies. It's also essential to create multinational regulations to oversee these efforts. By investing early in space-rated thorium MSRs, companies and agencies can position themselves at the forefront of a future energy economy that spans multiple planets.

Now that we've identified a technology baseline, we are ready for launch.

Step 2: The Moon Colony

The Vision

Imagine the Moon - silent, desolate, but not lifeless. In a crater near the lunar south pole, a steady light glows from a self-sustaining habitat. Beneath a meter of regolith shielding, a thorium MSR hums quietly, supplying warmth, light, and life. A nearby rover returns from an in-situ thorium deposit, soon to be processed into reactor fuel.

This vision echoes what Kirk Sorensen once described:

> *"Thorium is also common on the Moon and easy to find. Thorium has an electromagnetic signature that makes it easy to find, even from a spacecraft. With the energy generated from a liquid fluoride thorium reactor, we could recycle all of the air, water, and waste products within the lunar community. Doing so would be an absolute requirement for success. We could grow the crops needed to feed the members of the community even during the two-week lunar night, using light and power from the reactor. It seemed like the liquid fluoride thorium reactor, or LFTR, could be the power source that could make a self-sustainable lunar colony a reality."*

— Kirk Sorensen, TED Talk[1]

Inside the habitat, the crew wakes up to warm, filtered air and purified water. Greenhouse crops thrive under LED arrays warmed by low-grade reactor waste heat. Closed-loop systems manage CO_2 levels and recycle air and water to sustain a carbon-neutral biosphere. During the two-week lunar night, when temperatures plummet and sunlight vanishes, the reactor ensures the lights stay on and life endures. Outside, robots 3D-print new structures from sintered regolith while the AI-guided MSR manages neutron flux, thermal gradients, and system diagnostics.

Moon Base Powered by Thorium Molten Salt Reactors

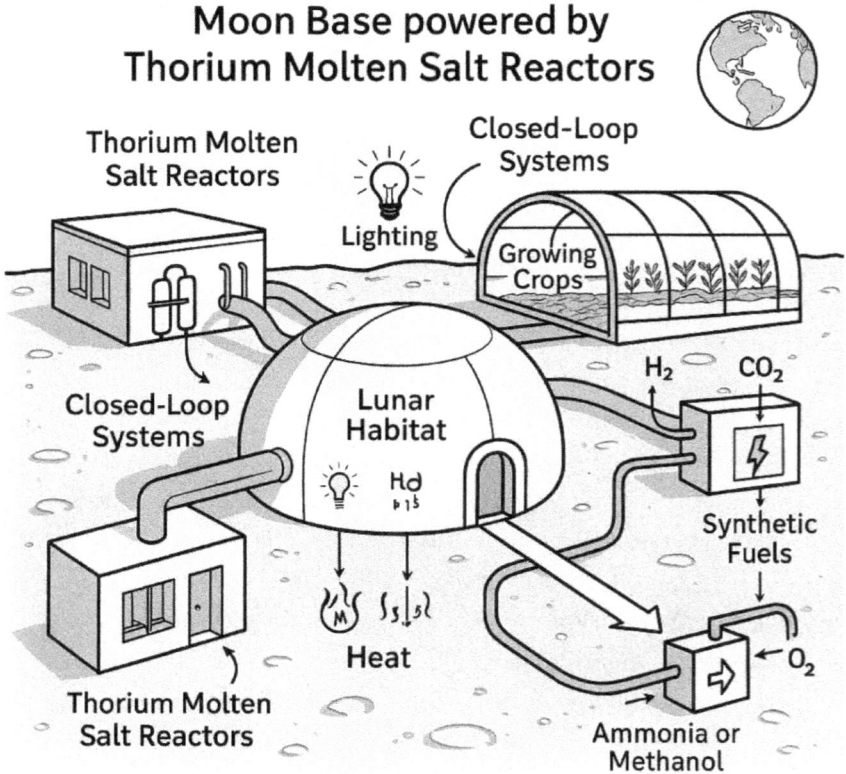

Moon Base powered by
Thorium Molten Salt Reactors

The Mission

Deploy thorium MSRs on the Moon to power long-duration lunar habitats. Demonstrate sustainable life support, in-situ fuel potential, and closed-loop environmental controls under continuous reactor operation.

Power Load Requirements

Lunar habitats require at least 40 kWe to maintain operations during the lunar night, including:

Lighting, air circulation, heating, and fluid systems for habitability

Greenhouse systems for crop production

Electrolysis units for oxygen generation and water reclamation

Communications arrays for Earth transmission

Robotic systems and 3D printers

Reactor control and monitoring systems

Scalable Space MSR: Scale to 40–100 kWe for long-duration lunar base operations.

Prefueled with thorium-232 and seeded with U-233, this compact reactor would deliver continuous heat and electricity. The system would be autonomous, radiation-hardened, and optimized for low-maintenance operation.

Relevant Space Tech Toolkit Capability

REACTOR AND FUEL SYSTEMS: KILOPOWER IS ACTIVE AND PROVIDES 1 TO 10 KILOWATTS. COPENHAGEN ATOMIC'S REACTORS AND NASA's FSP ARE STILL PROTOTYPES.

HEAT MANAGEMENT: PROJECT PELE'S HIGH-TEMPERATURE LOOPS REACHED PROTOTYPE STAGE.

AI AND CONTROLS: SpaceX's avionics and NASA's DART AI are currently active.

POWER SYSTEMS: Kilopower Stirling is active, while the Brayton cycle is still a prototype.

LAUNCH AND INTEGRATION: SpaceX Starship cargo systems are active.

NASA is actively developing compact fission surface power systems to deliver baseline energy for lunar operations. The initial focus is on HALEU-fueled systems, but molten salt and thorium fuel cycles are gaining attention.

India's Chandrayaan-1 mission confirmed thorium-rich regolith, making in-situ thorium utilization (ISRU) a future possibility.

The European Space Agency's Moon Village concept and a 2021 ESA-funded feasibility study have also examined molten salt thorium reactors for lunar deployment, highlighting thorium's thermal efficiency and abundance.

International consortia and private ventures are exploring high-temperature molten salt microreactors for space applications. NASA's Kilopower project, in partnership with the Department of Energy and Los Alamos National Laboratory, successfully demonstrated the viability of small fission reactors for space. Although current designs use uranium-235, the modular molten salt architecture offers a foundation for future thorium variants.

The Artemis program's[13] long-term vision includes the Artemis Base Camp, as outlined in NASA's 2020 "Plan for Sustained Lunar Exploration and Development" and the 2022 "Moon to Mars Objectives."

The plan calls for:

A crew habitat at the lunar south pole
40 kWe-class fission power systems
Pressurized and unpressurized rovers
Systems to extract water ice and generate oxygen
Integrated communications and scientific payloads

A dependable fission power system is critical for surviving the 14-day lunar night, when solar arrays are nonfunctional and batteries are insufficient.

Our concept design for a Scalable Space MSR would meet the Artemis Base Camp requirements. The next challenging step is to adapt the Scalable Space MSR into a propulsion system.

Step 3: The Journey to Mars: Advanced Nuclear Propulsion

The Vision

To bridge the vast and unforgiving distance between Earth and Mars, propulsion systems must advance beyond traditional chemical rockets. Two promising technologies are leading this new era.

Nuclear Thermal Propulsion (NTP) heats hydrogen propellant to over 2,500°C, allowing for high-thrust maneuvers with three times the efficiency of conventional rockets. Meanwhile, Nuclear Electric Propulsion (NEP), powered by a thorium MSR, generates continuous electricity for ion engines, enabling sustained acceleration over months with unparalleled fuel efficiency.

This spacecraft functions as a self-sustaining habitat. A rotating ring simulates Mars's gravity (0.38 g) and supports closed-loop life support systems, including electrolysis, CO_2 scrubbing, and water reclamation. At the ship's core lies the MSR, which channels heat through molten salt loops to radiator wings that glow as they dissipate thermal energy into space.

Lessons learned from lunar missions have equipped us to manage heat rejection, radiation shielding, and autonomous system control, capabilities that are now integrated into this craft designed for deep space exploration.

The Mission

Deploy a dual-mode, MSR-powered deep-space transport that combines NTP for high-thrust burns and NEP for continuous ion drive, aiming to complete Mars transit in under 150 days. The same reactor must power habitat life support, scientific labs, and communications while sustaining propulsion through varying mission phases.

Power Load Requirements

Subsystem Power Draw	(kWe)
Life Support, Habitat Ops	25–30
Artificial Gravity, Guidance	5
AI, Navigation, Data Systems	8–10
Science & Experimentation Labs	10
Communications (Laser + RF)	5–8
Ion Propulsion (NEP)	150–250
Thermal Pumps, Heat Transfer	3–5
Total Continuous Load	210–320

Scalable Space MSR: Scale to 250 kWe hybrid for NTP and NEP propulsion stack.

During NEP cruise, propulsion dominates the load. For rapid maneuvers or orbit insertion, NTP pulses are used independently, requiring short-term thermal management via reactor bypass systems.

Relevant Space Tech Toolkit Capability

REACTOR AND FUEL SYSTEMS: KILOPOWER IS OPERATIONAL AND PROVIDES 1 TO 10 KILOWATTS OF POWER. COPENHAGEN ATOMIC'S REACTOR IS IN THE PROTOTYPE STAGE, AND NASA'S FSP IS ALSO IN THE PROTOTYPE STAGE.

Heat Rejection and Thermal Management: Project Pele's high-temperature loops reached the prototype stage.

AI and Autonomous Controls: SpaceX's avionics and NASA's DART AI systems are operational.

Power Conversion Systems: The Kilopower Stirling system is operational, while the Brayton cycle is still in the prototype stage.

Launch and Integration: SpaceX Starship cargo systems are operational.

Notably, Russia's Zeus (Transport and Energy Module), under development by Roscosmos, aims to use a fission reactor to power Hall-effect thrusters for interplanetary missions[14]. Though still in the concept stage, Zeus has received funding and could evolve toward molten salt or hybrid designs[15]. In the U.S., NASA and DOE's Kilopower and Fission Surface Power projects are advancing compact fission reactors in the 1–40 kWe range. DARPA's DRACO program[16], led by BWXT[17] and Lockheed Martin[18], focused on NTP systems for cislunar operations, with potential extension to Mars-class missions.

Innovation Needed

To make the Scalable Space SMR design work for space travel, it must use a hybrid propulsion approach that combines two types of nuclear power: nuclear thermal propulsion and nuclear electric propulsion. NTP utilizes a fission reactor to heat hydrogen, providing high-thrust burns, while NEP uses the same reactor to generate electricity for ion engines during prolonged space missions.

The thorium-fueled MSR creates fissile uranium-233, which produces heat for the Brayton cycle. This provides the electricity needed to power onboard systems, including ion thrusters. In the NTP process, hydrogen is heated in the reactor core and then expelled through a nozzle to generate thrust. This hybrid system supports NTP thrust bursts using heated hydrogen and NEP through ion thrusters, combining the rapid maneuverability of thermal propulsion with the endurance of electric drive.

To realize this hybrid platform, six major advancements are required:

1. **Multi-Mode Reactor Throttling** – Enable reactors to switch between steady-state electric generation and pulsed high-thrust operations.
2. **Microgravity Brayton Cycle** – Turbines with magnetic bearings, non-lubricated seals, and low-gravity vibration damping must be validated in space.
3. **Radiation-Hardened Autonomy** – AI systems must manage all spacecraft operations with minimal Earth input, including power arbitration and failure prediction.
4. **Scalable Ion Propulsion** – Ion thrusters must evolve to megawatt-class power levels (vs. current 2–7 kWe) to utilize the full potential of MSRs.
5. **Thorium Fuel Logistics in Space** – Reactor salt chemistry and waste isolation must be adapted for microgravity, with robotic maintenance capability.
6. **Multifunctional Shielding** – Use mission-critical components (e.g., water tanks, food storage) as part of radiation shielding to reduce mass.

This NEP-NTP hybrid modification of the Scalable Space SMR is entirely fission-based.

Looking Ahead: Fission-Fusion Upgrade

A far-future propulsion alternative merges the strengths of fission and fusion into a single integrated system, offering the potential for deep space travel at unprecedented speeds and efficiencies. In this hybrid design, a compact thorium-based fission reactor provides continuous electrical power for life support, navigation, and fusion reactor ignition, while a fusion drive, such as a deuterium-tritium or deuterium-helium-3 system, delivers bursts of ultra-high specific impulse for fast interplanetary or even interstellar transit. The fusion stage provides the propulsion muscle, producing thrust through the controlled reaction of light nuclei, while the fission stage ensures steady baseline power and eliminates the need for massive solar arrays or chemical fuel reserves. Benefits include vastly reduced transit times to Mars and beyond, superior fuel efficiency, reduced radiation shielding needs (through smart directional exhaust), and a scalable architecture for robotic or crewed missions. This system, currently speculative, could redefine mission profiles across the solar system, enabling rapid return flights, dynamic rerouting, and the possibility of true off-world expansion.

Step 4: Mars: The Forge of New Worlds

The Vision

Mars is distinctly different from the Moon. It has an atmosphere composed of carbon dioxide, and although it receives sunlight, it also experiences dust storms that can last for weeks, obscuring solar energy.

Wind whips across the surface, scattering dust that clings to the solar panels, but the nuclear reactor operates without interruption. On Earth, a power outage is merely an inconvenience; on Mars, it can be a matter of life or death.

The energy demands here are substantial: 30 kW for life support, 60 kW for industry, and 20 kW for agriculture. At the heart of the settlement, two MSRs glow within basalt casings. One reactor powers an oxygen extraction facility, water systems, and a greenhouse dome, while the other operates a sintering furnace that transforms Martian dust into load-bearing bricks and tiles for roads and shelters.

Beneath a dome, spinach grows in CO_2-enriched air warmed by reactor waste heat. Outside, autonomous rovers 3D-print spare parts from locally sourced materials. This is not just about survival; it represents the farthest outpost of human existence.

The Mission

Establish dual-reactor hubs to support agriculture, regolith processing, and oxygen extraction on Mars. Validate the startup procedures of autonomous reactors and develop long-term safety protocols to ensure they function effectively under Martian conditions. Enhance manufacturing and self-sufficiency by utilizing thorium MSRs to power on-site production of spare parts, infrastructure, and radiation shielding. Integrate 3D printing, sintering, and in-situ resource utilization-fed fabrication lines for use on Mars and beyond.

The Power Load Requirements

Here, power demands are higher to include more industrial support. Establish a 120–150 kWe thorium MSR system to support a full-scale Martian outpost. This includes:

30 kWe for habitation systems
60 kWe for ISRU and regolith processing
20 kWe for greenhouses and hydroponics
Additional capacity for rovers, sensors, and comms

Scalable Space MSR: Scale to 120–150 kWe distributed across two 80 kWe units. The two 80 kWe MSRs provide built-in redundancy, ensuring the colony remains operational even during equipment servicing or unexpected events.

Relevant Space Tech Toolkit Capability

REACTOR AND FUEL SYSTEMS: CNSA 1 MWe initiative (Concept Design), ESA MSR studies (Concept Design), NASA FSP (Prototype stage).

HEAT REJECTION AND THERMAL MANAGEMENT: ESA heat transfer concepts (Concept Design).

AI and Autonomous Controls: NASA DART & SpaceX avionics (Operational).

Robotics: Boston Dynamics/Palladyne AI for mobility and handling (Operational).

In-Situ Manufacturing: NASA ISRU additive fabrication and sintering (Prototype stage).

Fuel Handling: Los Alamos salt chemistry research (Prototype stage).

China's CNSA has announced plans to develop a 1 MWe-class nuclear power system for lunar and Martian bases. This includes thorium MSRs in early-stage designs and aligns with the Chinese Academy of Sciences' focus on long-term, self-sustaining habitats.

The European Space Agency (ESA) has evaluated MSR concepts in its Moon and Mars initiatives, emphasizing modularity, transportability, and safety in radiation-rich, low-gravity environments.

Private firms, such as Oklo[19] and Ultra Safe Nuclear Corporation (USNC)[20], are advancing microreactor platforms originally designed for terrestrial deployment. These designs, while currently based on HALEU fuel, are being explored for Martian adaptation and may incorporate molten salt thorium variants in future iterations.

Innovation Needed

To realize a thorium-powered Martian settlement, we must bridge critical engineering gaps across multiple domains: reactor design, materials science, heat management, autonomous systems, and resource integration. The following innovations are essential to adapt MSR technology for the harsh, isolated, and resource-limited conditions of the Red Planet.

Reactor Miniaturization and Autonomy: MSRs must be compact enough for transport aboard vehicles like SpaceX's Starship and capable of autonomous startup and operation. This includes AI-managed systems, thermal regulation, and passive safety engineered for Martian conditions like CO_2 frost and abrasive dust.

Molten Salt Chemistry in Martian Conditions: Molten salt flow in 0.38g gravity remains untested. Engineers must model and validate flow dynamics, thermal expansion, and salt chemistry in reduced gravity. Materials like Hastelloy-N must resist both thermal cycling and Martian environmental wear. If thorium is to be sourced locally, ISRU-based extraction and processing systems must be autonomous and dust-resilient.

Compact Heat Rejection: Radiative cooling is the only viable method on Mars. Radiator panels must be high-emissivity, foldable, and self-cleaning to shed dust. Future systems may use electrostatic or phase-change tech to optimize heat dissipation in space.

Shielding and Habitat Design: With no protective magnetosphere, Mars requires robust radiation shielding. Sintered regolith bricks, underground reactor placement, and multipurpose shielding (like water tanks) must all be designed for robotic assembly prior to crew arrival.

MSR–ISRU Integration: The synergy between reactor waste heat and ISRU systems is critical. Thermal energy can power regolith-based oxygen extraction, brick manufacturing, and water electrolysis. Standardized interfaces are needed to couple reactors with these components in modular, serviceable architectures.

Building with Local Resources

Thorium-powered systems offer not only energy but infrastructure. With access to Martian CO_2, regolith, and ice, MSRs could power reactors that extract oxygen, metals, and water for use in construction, life support, and even synthetic fuel production. Structures, tools, spare parts, and shielding could be manufactured on-site, reducing launch mass and supporting a self-sufficient colony.

Synthetic Fuel Production

Using CO_2 from the Martian atmosphere and hydrogen from buried ice, reactors could enable local synthesis of fuels such as methane, ammonia, or methanol. These would power rovers, heat greenhouses, and even provide propellant for return flights, supporting full-cycle energy independence.

A Future Hybrid Fission-Fusion Solution?

Looking further ahead, a base could adopt a hybrid system: thorium MSRs providing a steady energy baseline, supplemented by compact fusion reactors for peak loads, such as intensive regolith processing, industrial-scale 3D printing, or expanded habitation during crew rotations. Both systems could share a common FLiBe loop for tritium breeding and thermal integration, enhancing redundancy and operational resilience. This hybridization would represent a major step toward energy self-sufficiency on Mars and beyond.

A Springboard to Deep Space

Mars is unforgiving of errors, leaving no margin for improvisation. This characteristic establishes it as the ultimate testing ground for innovation. The implementation of thorium MSRs on Mars represents not only an engineering challenge but also a bold declaration of our intent: to survive, adapt, and flourish as a base that nurtures new technologies while serving as a launching pad for further space exploration.

Step 5: Deep Space Reach: Miniature Reactors for Outer Planet Probes

The Vision

As humanity's ambitions extend beyond the inner solar system, our power systems must evolve accordingly. The Voyager probes ventured into the depths of space relying on diminishing plutonium heat. However, the next generation of deep space missions will require significantly more: greater power, more extended longevity, and enhanced capabilities. Miniature thorium MSRs, approximately the size of a barrel, can provide power to these probes for 30 to 50 years. This would enable them to support instruments capable of mapping subsurface oceans on Europa, analyzing hydrocarbons on Titan, or scanning exoplanet atmospheres from the far reaches of the Kuiper Belt. These probes will act as self-sustained outposts of intelligence, silent scouts that continuously stream scientific data and fill in the edges of our exploration map.

The Mission

Develop ultra-compact thorium MSRs and deployable heat rejection systems for long-duration scientific probes beyond Jupiter's orbit. Enable high-data-rate communications, sustained onboard science, and electric propulsion in regions where solar power is no longer viable.

The Power Requirement

Beyond the asteroid belt, solar irradiance drops below practical thresholds. Meanwhile, plutonium-238, used in RTGs, is rare and expensive. Miniaturized thorium MSRs offer a scalable, abundant alternative capable of providing 10–20 kWe continuously. This output could support:

> Long-range ion propulsion for station-keeping or orbital insertion
> Ground-penetrating radar for icy moon exploration
> Spectroscopy payloads for biosignature and exoplanet detection
> Quantum-encrypted communications or high-gain optical relays
> Radiothermal heating to protect systems in deep cold

Scalable Space MSR: Scale to 10–20 kWe ultra-miniaturized probe MSRs.

This steady, autonomous energy supply could extend scientific operations for decades, making possible multi-target missions that were previously unthinkable.

Relevant Space Tech Toolkit Capability

REACTOR AND FUEL SYSTEMS: THE KILOPOWER[2] STIRLING CYCLE IS CURRENTLY OPERATIONAL, DEMONSTRATING EFFECTIVE POWER GENERATION. ADDITIONALLY, DESIGN STUDIES FOR FISSION MICROREACTORS ARE UNDERWAY AS PART OF EARLY PATHFINDING EFFORTS IN ADVANCED NUCLEAR TECHNOLOGY.

HEAT REJECTION: COMPACT RADIATOR CONCEPTS THAT FOCUS ON EFFICIENT HEAT DISSIPATION, CURRENTLY AT THE CONCEPT DESIGN STAGE TO OPTIMIZE PERFORMANCE FOR FUTURE SYSTEMS.

POWER CONVERSION SYSTEMS: STIRLING SYSTEMS ARE OPERATIONAL AND EFFECTIVELY CONVERTING THERMAL ENERGY INTO ELECTRICAL POWER, SHOWCASING RELIABLE TECHNOLOGY FOR SPACE AND TERRESTRIAL APPLICATIONS.

AI AND AUTONOMY: AI DIAGNOSTICS SPECIFICALLY TAILORED FOR EARTH-BASED FISSION PLATFORMS ARE PRESENTLY IN THE PROTOTYPE STAGE, AIMING TO ENHANCE SAFETY AND OPERATIONAL EFFICIENCY.

Most current deep space probes use radioisotope thermoelectric generators (RTGs), which generate less than 300 watts of electricity and operate at low efficiency ($\sim$6%). NASA's Next-Generation RTG (NGRTG)[2] program aims to improve performance, but output remains limited.

The Kilopower project has demonstrated 1–10 kWe reactors with Stirling conversion, but not yet at the miniaturized scale needed for interplanetary probes. Concepts for fission reactors below 100 kg total system mass remain unproven.

Russia's Zeus TEM project envisions higher-power space reactors but targets crewed transport rather than microprobe power systems. No current mission has yet flown a molten salt reactor in space, although laboratory research into compact fuel salts and microgravity-compatible containment is underway.

Innovation Needed

To realize thorium MSR-powered interplanetary probes, key innovations must be achieved:

Reactor Miniaturization: Shrink reactor mass to under 200 kg including shielding, radiators, and power conversion. Innovations in salt composition, shielding geometry, and compact Brayton or Stirling cycles will be critical.

Autonomous Start-up and Self-Regulation: Design reactors that can self-initiate and operate without human intervention for decades. AI-driven fault detection, fuel burnup tracking, and self-repair systems are essential.

Microgravity-Compatible Radiators: Deploy lightweight, high-emissivity radiator fins that work in deep space. Must handle freeze-thaw cycles, dust impacts, and radiation exposure over decades.

Dust-Hardened Electronics and Shielding: Probes must shield against galactic cosmic rays, solar flares, and micrometeoroids, while minimizing mass. Consider layered shielding using structural materials, regolith-inspired composites, or multifunctional insulation.

Long-Term Propellant Compatibility: Match reactor output with ion or Hall-effect engines capable of operating for years. Require ultra-long-life thrusters, xenon or alternative propellant storage, and thermal management integration with reactor systems.

Probes will Map the Waystations

Future deep space probes will be unmanned and powered by thorium energy. Equipped with miniature MSRs, these intelligent machines will quietly and tirelessly explore the cosmos, mapping resources that can be mined as waystations for deeper space exploration.

Thorium's potential extends beyond reactors and resource resilience - it's now being explored for cutting-edge quantum technology. In 2025, DARPA launched the SUNSPOT program (Sources for Ultra-violet Nuclear Spectroscopy of Thorium) to develop a next-generation nuclear clock based on a unique property of the thorium-229 isotope. Unlike atomic clocks, which rely on electron transitions, this clock would harness transitions within the nucleus itself, enabling precision timing that is more stable, less sensitive to environmental noise, and resilient enough for deep-space navigation or secure communications. Thorium's nuclear structure is so well-suited to this purpose that researchers have focused specifically on its ultraviolet resonance near 148 nanometers; an elusive transition only now becoming accessible with advances in vacuum-ultraviolet laser technology. If successful, thorium could anchor an entirely new class of ultra-compact, radiation-hardened quantum timing systems. If researchers can tune a laser to this faint signal, they could create a tiny nuclear clock that stays accurate for decades, ideal for space probes, secure communications, and even GPS systems on the Moon or Mars. It's one more way thorium could quietly shape the future of space exploration, not just by powering our journey, but by keeping it perfectly on time.

Step 6: The Working Frontier: Mining and Refueling Waystations

The Vision

On a carbon-rich asteroid in the Main Belt, a drill hums under the low gravity. Nearby, a chemical reactor splits carbon dioxide (CO_2) and hydrogen into methane fuel. Ice is harvested and then electrolyzed to produce oxygen and hydrogen. Metals are smelted from regolith. The power source for all this? A thorium molten salt reactor.

These facilities serve as the logistical backbone of interplanetary exploration. With thorium's stable and long-lasting energy output, these autonomous stations can operate as ISRU hubs, fuel depots, and resource factories. They provide propellant, oxygen, and raw materials to vessels traveling to Mars, the outer planets, or beyond.

Thorium decentralizes energy access, allowing each station or colony to function independently of Earth. This promotes international collaboration rather than competition. In a domain where no nation can lay claim to territory according to the principle established in the Outer Space Treaty of 1967, thorium-powered infrastructure levels the playing field, ensuring that all stakeholders have access to energy and resources.

From the shadow of Ceres to the orbit of Vesta, autonomous mining stations powered by compact MSRs will extract water, break down hydrocarbons, and refine metals. These modular installations will grow incrementally, enabled not only by thorium power but also by artificial intelligence, robotics, and additive manufacturing. Some will be crewed, while most will operate autonomously, powered by thorium cores and functioning with near-complete independence.

Today's robotic platforms, such as those developed by SpaceX or Boston Dynamics for use on Earth, could be adapted for microgravity excavation and ISRU tasks. What currently mows lawns and inspects launch pads may soon mine regolith and construct fuel tanks in space.

The Mission

Demonstrate thorium MSR-powered in-space mining operations and establish ISRU-driven refueling depots on asteroids and small planetary bodies. Prove the viability of autonomous, energy-independent infrastructure for fuel synthesis, metal extraction, and life support resupply.

The Power Requirement

Mining stations and ISRU platforms will require modular power systems scaled from 20 to 200 kWe, depending on activity:

20–40 kWe for basic robotic excavation, lighting, and communication relays
40–80 kWe for water electrolysis, CO_2 cracking, and cryogenic fuel storage
100–200 kWe for integrated operations with metal refining, additive manufacturing, and docking services for visiting spacecraft

Scalable Space MSR: Scale to 20–200 kWe scalable modular systems.

Each thorium MSR module must be compact, durable, and scalable, delivering uninterrupted power in harsh thermal and radiation environments.

Relevant Space Tech Toolkit Capability

REACTOR AND FUEL SYSTEMS: COPENHAGEN ATOMIC'S CONTAINERIZED CONCEPTS ARE IN THE PROTOTYPE STAGE.

HEAT REJECTION AND THERMAL MANAGEMENT: ESA IS DESIGNING MODULAR LOOPS AS A CONCEPT.

AI AND AUTONOMOUS CONTROLS: SpaceX AND DART AVIONICS ARE CURRENTLY IN OPERATION.

ROBOTICS AND ISRU: BOSTON DYNAMICS AND NASA ARE WORKING ON THE VIPER PROJECT, WHICH IS IN THE PROTOTYPE STAGE.

IN-SITU MANUFACTURING: NASA IS USING ADDITIVE FABRICATION, CURRENTLY IN THE PROTOTYPE STAGE.

FUEL HANDLING: LOS ALAMOS IS CONDUCTING CHEMISTRY RESEARCH AT THE PROTOTYPE STAGE.

Enabling Advancements Applicable for Space Mining

Many of today's technologies will mature and enable future space mining missions.

NASA's Lunar ISRU and VIPER Rover: Testing regolith mining and volatile extraction on the Moon

ESA's PROSPECT program: Volatile prospecting tools designed for autonomous drilling

SpaceX, Boston Dynamics, Anduril, and Palladyne AI Robotics: Terrestrial platforms with AI and mobility features that can be repurposed for remote space environments

Kilopower (NASA/DOE) and Copenhagen Atomics (commercial): Compact fission systems in development that serve as templates for scalable, thorium-capable systems

No mission to date has demonstrated ISRU at scale or integrated fission energy with mining operations in space, but the component technologies are converging.

Innovation Needed

Modular, Low-Mass MSRs: Thorium MSRs must be miniaturized for launch and surface deployment. Target reactor mass: under 500 kg, with integrated shielding and autonomous startup capabilities.

Fully Autonomous ISRU Systems: Mining and fuel synthesis must be AI-managed with limited human intervention. Autonomous robotics, error correction algorithms, and machine learning systems must coordinate excavation, processing, and maintenance.

Dust- and Radiation-Hardened Hardware: Asteroid environments are abrasive, cold, and irradiated. Power systems, pumps, seals, and electronics must be ruggedized for decades of operation without crewed service.

Energy-Integrated Chemical Reactors: CO_2 cracking, water electrolysis, and methane synthesis units must be thermally integrated with reactor waste heat for optimal efficiency. This requires standardized heat interfaces and real-time thermal balancing.

On-Site Additive Manufacturing: To achieve full logistical independence, ISRU systems must produce spare parts, tools, and replacement components via robotic 3D printing. This closes the supply chain loop and supports long-term mission sustainability.

Fueling the Exploration

The space economy will depend on logistics fueled by thorium. These mining and refueling outposts will enable humanity to expand throughout the solar system by harvesting, processing, and refueling with reactors. In this new frontier, thorium will serve as a key resource, supporting independence and powering waystations for our next big leap.

Step 7: Becoming Autonomous: New Tools for a New Era

The Vision

To move beyond Earth's orbit and stay there, our systems must become not just sustainable but intelligent. Autonomy is the key to permanence. In the harsh, remote environments of the Moon, Mars, asteroids, and deep space, the margin for error is razor thin. Real-time communication with Earth becomes infeasible due to delays. Crewed maintenance is limited by mass, cost, and survival constraints.

To meet these demands, we must equip reactors and support infrastructure with AI-driven autonomy, predictive diagnostics, and self-repair capabilities. Picture an MSR that senses microfractures forming in its containment wall, diverts flow, initiates robotic patch-work, and notifies Earth only when it's already resolved. Imagine a robotic 3D printing arms using sintered regolith to fabricate new shielding panels, drone swarms maintaining radiator arrays, and machine learning systems optimizing reactor output in real-time to match fluctuating loads.

Thorium MSRs will serve as more than power plants. Their high-grade heat can distill water, split molecules, produce breathable oxygen, and sterilize medical equipment. With the right systems in place, they could melt Martian ice, support atmospheric processing, and potentially aid in cryogenic stasis systems for crew preservation on interstellar journeys.

Autonomy extends to communications as well. Future data relays, especially laser-based optical links, will require 10-20 kW per node to maintain high-bandwidth connections across millions of kilometers. Only reactors can sustain the continuous, high-output energy that these systems demand.

The Mission

Design and deploy fully autonomous reactor control and mainte-nance systems that enable distant outposts to operate for years or decades without direct human oversight. These tools will serve as the operational backbone for AI-managed, self-repairing habitats and robotic scientific stations.

The Power Requirement

Autonomous systems require a continuous power supply across three tiers:

Reactor Core Operation: 20–50 kWe for AI controls, cooling pumps, diagnostics, and reactivity control

Peripheral Robotics: 10–30 kWe for 3D printers, manipulators, drones, and repair arms

Communications & Data Systems: 10–20 kWe for laser relays, satellite links, and onboard computing

Scalable Space MSR: Scale to 50–100 kWe per autonomous hub.

Scalable Space MSR: A single thorium MSR delivering 50–100 kWe can support an entire autonomous node with margin for redundancy and expansion.

Relevant Space Tech Toolkit Capabilities

AI AND AUTONOMOUS CONTROLS: KILOPOWER STARTUP PROTOCOLS (OPERATIONAL), SPACEX STARSHIP AI, (OPERATIONAL)

ROBOTICS: BOSTON DYNAMICS, PALLADYNE AI, ETH ZURICH MOBILITY PLATFORMS (OPERATIONAL), TESLA OPTIMUS MARS ANNOUNCEMENT (2025) (PROTOTYPE).

IN-SITU MANUFACTURING: NASA REGOLITH 3D PRINTING AND SINTERING (PROTOTYPE STAGE).

POWER CONVERSION: KILOPOWER (OPERATIONAL), BRAYTON TURBINES (PROTOTYPE STAGE).

COMMUNICATIONS SYSTEMS: HIGH-ENERGY LASER RELAYS (EARLY PATHFINDING).

Enabling Advancements Applicable for Autonomy

The technology enablers for autonomy can be found across industries. Here are some noteworthy developments:

NASA's VIPER and DART AI navigation systems: Demonstrate autonomous terrain navigation and obstacle avoidance

SpaceX Starship avionics: Built on real-time fault recovery and deep learning models

Boston Dynamics, Palladyne AI, and ETH Zurich robotics: Provide advanced manipulators and AI-guided mobility platforms

Kilopower and Fission Surface Power (FSP): Demonstrate modular, unattended reactor startup and shutdown protocols

These elements provide a robust foundation for building intelligent reactor systems that can detect, adapt, and recover from failure.

Innovation Needed

AI-Driven Reactor Control: Machine learning models must manage reactivity, temperature gradients, and neutron flux in real time. Algorithms need to process massive sensor datasets with zero-latency local decision-making.

Predictive Diagnostics: Systems must forecast failures before they happen, tracking corrosion, thermal fatigue, radiation damage, and coolant composition with high fidelity.

Robotic Self-Maintenance: Onboard tools should include robotic arms, microdrones, and modular repair kits. These systems must remove, print, and replace components on the fly without human assistance.

Autonomous Fabrication: Advanced 3D printing with local materials (e.g., regolith) will support part replacement, shielding upgrades, and thermal component replication, backed by AI-validated design libraries.

Secure and Resilient Communication: Autonomous nodes must use encrypted, fault-tolerant protocols to relay data, updates, and alerts to mission control across vast distances, even when Earth is temporarily out of reach.

Programmable Power: Imagine a future where nuclear materials can be turned up or down like a dimmer switch: decaying when we need energy, and quiet when we don't. That's the vision behind DARPA's Decay on Demand program. Announced in 2024, this research effort explores whether it's possible to speed up or control radioactive decay using precisely tuned beams of X-rays or other energy inputs. If successful, this could transform how reactors manage fuel and waste, especially in space. Instead of carrying large, passive isotope batteries, future missions could use compact fuel sources that release energy only when triggered. It could even allow reactors to reshape waste into safer or more useful forms on the fly. This level of control, deciding when and how fast a radioactive material decays, would be a game-changer, giving advanced thorium systems a new layer of intelligence. It means a system could not only generate power but adjust its fuel behavior based on mission demands, emergencies, or even habitat conditions. In the future, energy in space won't just be powerful; it could be programmable.

Autonomy for a Multiplanetary Future

Autonomy is vital for long term settlement in space. We require fully autonomous reactor control and maintenance systems capable of operating for years without human intervention. These systems will support AI-managed habitats and robotic stations. As humanity ventures beyond Earth, thorium-powered machines will not only function effectively but also adapt and self-repair, becoming essential for a multiplanetary civilization.

Technology Maturity Assessment

OPERATIONAL TECHNO-LOGIES

1. Molten salt reactor chemistry and material science, proven in terrestrial labs and testbeds.

2. Kilopower's successful demonstration of compact, low-power space fission units.

3. AI-driven sensors for fault detection and predictive maintenance, already in use on Earth and in space avionics.

NEAR-TERM DEVELOP-MENTS

1. Lunar greenhouses and closed-loop life support, advancing through Artemis-era test programs.

2. Martian regolith sintering for construction, demonstrated in analog environments and robotic studies.

3. High-efficiency radiators and Stirling or Brayton-cycle systems, which are under prototyping in programs like NASA FSP and DARPA Project Pele.

SPECULATIVE CONCEPTS

1. Cryostasis systems for long-duration human travel.

2. Terraforming principles, like atmospheric thickening or greenhouse warming.

3. Quantum communication networks, which are emerging from early terrestrial quantum encryption platforms.

Step 8: Peace Through Power: Thorium as a Catalyst for Collaboration in Space

The Vision

Throughout history, wherever scarcity has taken root, conflict has often followed. In the harsh frontiers of space, where life depends on continuous access to air, water, food, and power, the risks of resource-driven tension escalate. Yet, thorium molten salt reactors offer a unique opportunity to invert that dynamic. By decentralizing energy production, thorium enables every base, outpost, or vessel to operate independently, no longer reliant on Earth-bound fuel deliveries or vulnerable to energy monopolies.

Imagine colonies where energy abundance fosters diplomacy. Where settlements are not strategic vulnerabilities but resilient, autonomous communities. This vision requires more than technology; it demands policy.

The Artemis Accords: The Foundation

In the real world, the Artemis Accords[13], initiated by NASA in 2020, form a set of multilateral agreements that guide peaceful exploration and use of outer space. Signed by more than 30 countries, the accords promote principles such as transparency, interoperability, the sharing of scientific data, peaceful conflict resolution, and the responsible use of space resources. Critically, they emphasize cooperation over territorial control.

The Artemis Accords lay the groundwork for international collaboration, focusing primarily on surface activities, scientific exploration, and broad norms of behavior. What they do not yet address in detail is the governance of shared nuclear infrastructure, particularly the management, safety, and ethical use of high-power systems, such as MSRs.

THE THORIUM PROTOCOLS:
A TRANSFORMATIVE ADDITION

To extend the spirit of the Artemis Accords into the nuclear age of space exploration, we propose the establishment of the Thorium Protocols as a hypothetical but vital framework that would:

1. Establish reactor transparency and inspection regimes, ensuring all thorium MSRs operate under shared safety and environmental standards.

2. Enable emergency power-sharing agreements, allowing any base in crisis to access wattage from neighboring reactors, much like a shared electrical grid.

3. Set limits on territorial reactor claims, preventing any single nation or corporation from hoarding energy access in key resource zones.

4. Create conflict resolution mechanisms based on energy arbitration, where disputes are negotiated through energy-sharing agreements rather than territorial posturing.

From Scarcity to Solidarity

By removing the competition for fuel and solar access, thorium MSRs foster a new ethical paradigm, one where cooperation replaces conquest. Independent power generation no longer means isolation; it becomes the foundation of mutual resilience. In this world, energy is not a weapon. It is a guarantee of peace.

In the same way the Antarctic Treaty demilitarized the South Pole, the Thorium Protocols could de-weaponize energy in space. By promoting open access, reciprocal agreements, and cross-national oversight, these protocols could ensure that power systems become symbols of shared purpose, not domination.

The Mission

Foster international collaboration by sharing thorium energy systems. Turn power into diplomacy, ensuring that energy independence enables a cooperative, multinational human presence in space.

Power Load Requirements

Scalable Space MSR: 40–100 kWe shared infrastructure units per base.

Relevant Space Tech Toolkit Capability

GOVERNANCE FRAMEWORKS: THE ARTEMIS ACCORDS ARE CURRENTLY IN USE, WHILE THE THORIUM PROTOCOLS ARE A NEW AND PLAUSIBLE INTRODUCTION, BUT NOT YET IMPLEMENTED.

REACTOR AND FUEL SYSTEMS: KILOPOWER IS A WORKING SYSTEMS, WHILE COPENHAGEN ATOMIC'S REACTOR AND THE FSP IS IN THE PROTOTYPE STAGE.

AI AUTONOMY: KILOPOWER, DART, AND SPACEX HAVE OPERATIONAL AI SYSTEMS.

FUEL HANDLING & SAFETY OVERSIGHT: ORNL STANDARDS AND RESEARCH FROM LOS ` ARE IN THE PROTOTYPE STAGE.

INTEROPERABILITY ARCHITECTURE: THE SHARED LOGISTICS PROTOCOLS USED IN ARTEMIS-ERA PLANNING ARE IN THE PROTOTYPE STAGE.

Shaping a Collaborative Power Dynamic

Power has always shaped geopolitics. In space, it will shape civilization. With thorium as a steady flame and the Thorium Protocols as its guiding law, we can ensure that energy becomes the great equalizer, not the next great conflict. Peace through power is a system design, and a collaborative approach for how we build a future worth living in, together.

International Partnerships in Nuclear Space Energy Development

Country / Region	Lead Agencies / Companies	Key Projects / Collaborations	Relevance to MSRs / Thorium
United States	NASA, DOE, DARPA, Westinghouse	Kilopower Project, Fission Surface Power, DARPA DRACO, Project Pele	Potential for small MSRs for lunar outposts and spacecraft propulsion
European Union (ESA)	ESA, CEA (France), Rolls-Royce (UK), Tractebel (Belgium)	Lunar power feasibility studies, nuclear microreactor design for ESA missions	Evaluating molten salt and solid-fuel space reactor options
China	CNSA, China Academy of Space Technology (CAST), SINAP	Space reactor roadmap, lunar base planning, 2030s nuclear space tug	Prototype MSR (TMSR-LF1) could be adapted for orbital power
Russia	Roscosmos, Rosatom	TEM "Zeus" space tug using nuclear electric propulsion	Fast reactors tested; thorium-compatible R&D in place
India	ISRO, BARC, Department of Atomic Energy	Moon and Mars missions with nuclear auxiliary power in concept stage	Long-term thorium expertise via AHWR may transfer to space platforms
Japan	JAXA, Mitsubishi Heavy Industries	Feasibility studies on nuclear thermal propulsion, space microreactors	Focus on compact and safe reactor design integration
South Korea	KARI, KAERI	Collaborating with US on SMRs and nuclear R&D for space	Thorium MSRs considered under long-range lunar power scenarios
Canada	Canadian Nuclear Labs, NB Power, Moltex	Regulatory leadership in SMRs; exploring off-grid deployment models	Salt reactor R&D relevant to off-planet systems
United Arab Emirates	ENEC, space partnerships via MBRSC	Potential for nuclear-enabled desalination + lunar interest through Artemis Accords	UAE's nuclear experience provides regulatory precedent

1. Kirk Sorensen, ted talk, 'thorium: an alternative nuclear fuel.'

2. Nasa kilopower project: https://www.nasa.gov/directorates/spacetech/kilopower

3. Terrell, jeff. "copenhagen atomics: the danish startup building mass-produced thorium reactors." Medium, march 5, 2023.

4. Nasa fission surface power system overview: https://www.nasa.gov/press-release/nasa-announces-fission-surface-power-project-awards

5. Cnsa mars reactor announcement: chinese academy of sciences bulletin, 2022

6. Esa thorium msr feasibility studies: european space agency publications, 2021

7. Los alamos national laboratory msr salt chemistry research: https://www.lanl.gov

8. Oak ridge msr history: ornl technical memoranda, 1965–1975

9. Anduril autonomous systems: www.anduril.com

10. Boston dynamics robotics: www.bostondynamics.com

11. Tesla optimus robot announcement: https://en.wikipedia.org/wiki/optimus_(robot)

12. Nasa viper rover: https://www.nasa.gov/viper

13. Nasa artemis plan: 2020 plan for sustained lunar exploration and development

14. Anatoly Zak, "Russia's Nuclear Space Tug," russianspaceweb.com, updated February 2024, https://www.russianspaceweb.com/tem.html.;

15. TASS Russian News Agency, "Roscosmos Develops Space Tug with Nuclear Reactor," TASS, April 13, 2021, https://tass.com/science/1277477.

16. DARPA, "Demonstration Rocket for Agile Cislunar Operations (DRACO)," DARPA News, last modified July 26, 2023, https://www.darpa.mil/news-events/2023-07-26.

17. BWX Technologies, "BWXT to Build Nuclear Thermal Propulsion Reactor for DARPA's DRACO," BWXT Newsroom, July 26, 2023, https://www.bwxt.com/news/2023/07/26/bwxt-to-build-nuclear-thermal-propulsion-reactor-for-darpa-s-draco.

18. Lockheed Martin, "Lockheed Martin Selected for DRACO Spacecraft Design," lockheedmartin.com, July 2023, https://www.lockheedmartin.com/en-us/news/features/2023/draco-spacecraft.html.

19. Jeff Foust, "Oklo Seeks to Expand Microreactor Applications to Space and Lunar Surface," SpaceNews, April 19, 2023, https://spacenews.com/oklo-seeks-to-expand-microreactor-applications-to-space-and-lunar-surface/.

20. USNC-Tech, "USNC-Tech and NASA Collaborate on Space Nuclear Power Systems," Ultra Safe Nuclear Newsroom, December 14, 2022, https://www.usnc.com/news/usnc-tech-and-nasa-collaborate-on-space-nuclear-power-systems.

Acknowledgments

Rav Astra – Cover art, figure graphics, and design of the https://inov8r.com website

Allison Smith – Copy editing and line editing

Stella Joy – Line editing and proofreading

INOV8R Press LLC – Publisher

www.ingramcontent.com/pod-product-compliance
Lightning Source LLC
Chambersburg PA
CBHW040847210326
41597CB00029B/4755